Optimising Pesticide Use

Wiley Series in Agrochemicals and Plant Protection

Series Editors:

Terry Roberts, *Consultant, Anglesey, UK.*
Junshi Miyamoto (deceased), *Formerly of Sumitomo Chemical Ltd, Japan*

Previous Titles in the Wiley Series in Agrochemicals and Plant Protection:

The Methyl Bromide Issue (1996), ISBN 0 471 95521 3.
Edited by C. H. Bell, N. Price *and* B. Chakrabarti
Pesticide Remediation in Soils and Water (1998), ISBN 0 471 96805 6.
Edited by P. Kearney *and* T. R. Roberts
Chirality in Agrochemicals (1998), ISBN 0 471 98121 4.
Edited by N. Kurihara *and* J. Miyamoto
Fungicidal Activity (1998), ISBN 0 471 96806 4.
Edited by D. Hutson *and* J. Miyamoto
Metabolism of Agrochemicals in Plants (2000), ISBN 0 471 80150 X.
Edited by Terry R. Roberts

Forthcoming Title in the Wiley Series in Agrochemicals and Plant Protection:

Pesticide Residues in Food and Drinking Water *Edited by* Denis Hamilton *and* Stephen Crossley
Optimizing Pesiticide Use *Edited by* Michael F. Wilson

Optimising Pesticide Use

Edited by
MICHAEL F. WILSON
Central Science Laboratory, Sand Hutton, York, UK

WILEY

Telephone (+44) 1243 779777

Email (for orders and customer service enquiries): cs-books@wiley.co.uk
Visit our Home Page on www.wileyeurope.com or www.wiley.com

Other Wiley Editorial Offices

John Wiley & Sons Inc., 111 River Street, Hoboken, NJ 07030, USA

Jossey-Bass, 989 Market Street, San Francisco, CA 94103-1741, USA

Wiley-VCH Verlag GmbH, Boschstr. 12, D-69469 Weinheim, Germany

John Wiley & Sons Australia Ltd, 33 Park Road, Milton, Queensland 4064, Australia

John Wiley & Sons (Asia) Pte Ltd, 2 Clementi Loop #02-01, Jin Xing Distripark, Singapore 129809

John Wiley & Sons Canada Ltd, 22 Worcester Road, Etobicoke, Ontario, Canada M9W 1L1

Wiley also publishes its books in a variety of electronic formats. Some content that appears in print may not be available in electronic books.

Library of Congress Cataloging-in-Publication Data

Optimising pesticide use / edited by Michael F. Wilson.
p. cm. – (Agrochemicals and plant protection)
Includes bibliographical references and index.
ISBN 0-471-49075-X (pbk. : alk. paper)
1. Pesticides. I. Wilson, Michael F. II. Series.

SB951.O66 2003
632′.95 – dc21

2003047962

British Library Cataloguing in Publication Data

A catalogue record for this book is available from the British Library

ISBN 0-471-49075-X

Typeset in 10/12pt Times by Laserwords Private Limited, Chennai, India
Printed and bound in Great Britain by TJ International, Padstow, Cornwall
This book is printed on acid-free paper responsibly manufactured from sustainable forestry in which at least two trees are planted for each one used for paper production.

Contents

Contributors

D. M. Armitage
Central Science Laboratory, Sand Hutton, York, YO41 1LZ, UK

R. Bateman
IPARC and CABI Bioscience, Ascot, Berks., UK

C. H. Bell
Central Science Laboratory, Sand Hutton, York, YO41 1LZ, UK

B. R. Champ
'Carawah', RMB2, Read Road, Sutton, NSW 2620, Australia

A. J. Gilbert
Central Science Laboratory, Sand Hutton, York, YO41 1LZ, UK

G. A. Matthews
Imperial College of Science, Technology and Medicine, Silwood Park, Ascot, Berkshire, SL5 7PY, UK

P. C. H. Miller
Silsoe Research Institute, Wrest Park, Silsoe, Bedford, MK45 4HS, UK

C. S. Parkin
Silsoe Research Institute, Wrest Park, Silsoe, Bedford, MK45 4HS, UK

R. Sohm
Syngenta Crop Protection, Münchwilen, CH-4333, Switzerland

M. R. Thomas
Pesticide Usage Survey, Central Science Laboratory, Sand Hutton, York, YO41 1LZ, UK

J. C. van de Zande
Institute of Agricultural and Environmental Engineering, IMAG, P.O. Box 43, 6700AA Wageningen, The Netherlands

M. F. Wilson
Central Science Laboratory, Sand Hutton, York, YO41 1LZ, UK

N. Woods
Director, The Centre for Pesticide Application and Safety (CPAS), The University of Queensland, Gatton Campus, Queensland 4343, Australia

Series Preface

There have been tremendous advances in many areas of research directed towards improving the quantity and quality of food and fibre by chemical and other means. This has been at a time of increasing concern for the protection of the environment, and our understanding of the environmental impact of agrochemicals has also increased and become more sophisticated thanks to multidisciplinary approaches.

Wiley has recognized the opportunity for the introduction of a series of books within the theme 'Agrochemicals and Plant Protection' with a wide scope that will include chemistry, biology and biotechnology in the broadest sense. This series is effectively a replacement for the successful 'Progress in Pesticide Biochemistry and Toxicology' edited by Hutson and Roberts that has run to nine volumes. In addition, it complements the international journals *Pesticide Science* and *Journal of the Science of Food and Agriculture* published by Wiley on behalf of the Society of Chemical Industry.

Volumes already published in the series cover a wide range of topics including environmental behaviour, plant metabolism, chirality, and a volume devoted to fungicidal activity. In addition, subsequent topics for 2003–2004 include optimization of pesticide use, operator exposure and dietary risk assessment. These together cover a wide scope and form a highly collectable series of books within the constantly evolving science of plant protection.

As I write this preface, I am deeply saddened by the recent death of Dr Junshi Miyamoto, who contributed so much to this series as my Co-Editor-in-Chief. More significantly, Junshi will be remembered for his lifetime achievements in agrochemical biochemistry, toxicology and metabolism – and not least for the energy he displayed in international activities aimed at harmonizing knowledge within the field of agrochemicals. He leaves us with a wealth of scientific publications.

Terry Roberts
Anglesey
July 2003

THE SERIES EDITORS

Dr Terry R Roberts is an independent consultant, based in Anglesey, North Wales. He was Director of Scientific Affairs at JSC International based in Harrogate, UK from 1996 to **2002**, where he provided scientific and regulatory consulting services to the agrochemical, biocides and related industries with an emphasis on EU registrations.

From 1990 to 1996 Dr Roberts was Director of Agrochemical and Environmental Services with Corning Hazleton (now Covance) and was with Shell Research Ltd for the previous 20 years.

He has been active in international scientific organizations, notably OECD, IUPAC and ECPA, over the past 30 years. He has published extensively and is now Editor-in-Chief of the *Wiley Series in Agrochemicals and Plant Protection*.

Dr Junshi Miyamoto (deceased) was Corporate Advisor to the Sumitomo Chemical Company, for 45 years, since graduating from the Department of Chemistry, Faculty of Science, Kyoto University. After a lifetime of working in the chemical industry, Dr Miyamoto acquired a wealth of knowledge in all aspects of mode of action, metabolism and toxicology of agrochemicals and industrial chemicals. He was a Director General of Takarazuka Research Centre of the Company covering the areas of agrochemicals, biotechnology as well as environmental health sciences. He was latterly Present of Division of Chemistry and the Environment, IUPAC, and in 1985 received the Burdick Jackson International Award in Pesticide Chemistry from the American Chemical Society and in 1995, the Award of the Distinguished Contribution of Science from the Japanese Government. Dr Miyamoto published over 190 original papers and 50 books in pesticide science, and was on the editorial board of several international journals including *Pesticide Science*.

Preface

Pest control and thus a reliance on pesticides, has been around for a long time. Records show that the Ancient Egyptians used alkaloid containing hemlock and aconite and Homer describes how Odysseus "fumigated the hall, house and court with sulphur to control pests". The amount of effort man has put into pest control is often inversely proportional to the food supply. In times of famine, we are less keen on sharing the food supply with other species – pests – or tolerant of spoilage to our food from pest activity. Conversely, in times of plenty, as is currently the case in the developed world, people have the option of seeking food grown without chemical pest control and balancing the possible risks from chemical residues with the perceived benefits, quality and abundance of untreated foods. This is not a global option.

World Wars I and II served as a watershed for the modern chemical industry. Many chemicals and technologies initially developed for warfare, were later adapted, developed further and focused on civilian uses. Swiss chemist Paul Müller discovered the insecticidal properties of DDT in 1939, an innovation that later earned him the Nobel Prize. German scientists experimenting with nerve gas during World War II synthesized the organophosphorus insecticide parathion, marketed in 1943, and still in use today. Throughout the 1950s and 60s, these types of chemicals became major pest control agents.

However, "Silent Spring", Rachel Carson's challenge to the perceived abuse of synthetic pesticides, published in 1962, initiated the a movement toward strict agrochemical regulation and the far more rigorous assessment of risks and benefits from pesticide use.

Whilst today's pesticides are designed to have fewer residues at harvest, persist in the environment for shorter periods and are to be less lethal to non-target species than the early days of calcium arsenate and DDT, the chemistry and biochemistry of pesticide design continues to improve. Such improvements are not confined to the chemistry and biochemistry of the active ingredients of pesticides. The way in which an active ingredient is formulated into a usable preparation and the engineering behind the manner in which it is applied to the crop have an equal part to play in minimising the risks from pest control. Each activity is designed to ensure that the pesticide molecule reaches and remains on its target – and preferably only its target – by adjusting the physio-chemical properties of the formulation and the precision of the application methods. Finally, pesticide use is part of broader agricultural practice and the non-prophylactic, limited use of pesticides as part of integrated pest management programmes is often the optimal way of achieving adequate pest control with minimal chemical use.

As editor of this volume, I am grateful to the authors that have contributed to this work. It is hopefully a fresh look at possible means of optimising pesticide use. Measuring precisely what is being used in response to real pest pressure and improving the application technologies are as important as the design and development of new chemistries. These facets are often overlooked and I hope this volume partly redresses this imbalance. I also would like to thank colleagues at the Central Science Laboratory for their help and advice during the production of manuscripts and for their patience during the long gestation period. I am particularly grateful to Andrew Gilbert for both his contribution and his help in collating some of the other chapters. Finally, I would like to thanks the series editors and staff at John Wiley & Sons, Ltd for useful comments and suggestions through the production of this volume.

MICHAEL WILSON

York
June 2003

1 Optimising Pesticide Use – Introduction

M.F. WILSON
Central Science Laboratory, Sand Hutton, York, YO41 1LZ, UK

SUMMARY

'Optimising pesticide use' is a very broad phrase that can be interpreted in a number of ways. At one extreme, it can imply the minimisation and prudent use of pesticides in agriculture and horticulture, but it can also be interpreted as the complete elimination of their use. However, it is probable, if not certain, that pesticides will continue to play a vital part in the safe and economic production of food in the foreseeable future. Notably, outside the developed world, pest-control strategies, including the use of chemicals, are essential for adequate food production and for current human health strategies.

This book, in recognising that pesticide use will occur, seeks to bring together the wide range of scientific disciplines necessary to ensure best practice, through monitoring what is used and improving how it is formulated and applied. The science of 'improving' or developing new chemistries as a means of optimising pesticide use is also touched upon. However, this is a subject in its own right, and best left to more specialist publications.

This book does cover in-depth the monitoring of usage, developments in application technology, including specialist and aerial application, handling and dose control and formulating pesticides. In addition, rational pesticide use (RPU) is a development of integrated pest management techniques, and, as such, must be considered as a central part of any consideration of optimisation. Finally, this publication concludes with a discussion of complementary pest control methods.

The interaction between the needs of agriculture, environmental protection and concerns for human health is complex, and is depicted in Figure 1.1. Human health implications cannot be simply regarded as negative and restricted to the potential risks associated with the presence of pesticide residues in foods. It is also essential to take into account the undoubted benefits from the control of infectious diseases vectored by pests and the contribution to health from a varied and safe food supply.

Optimising Pesticide Use Edited by M. Wilson
 ISBN: 0-471-49075-X

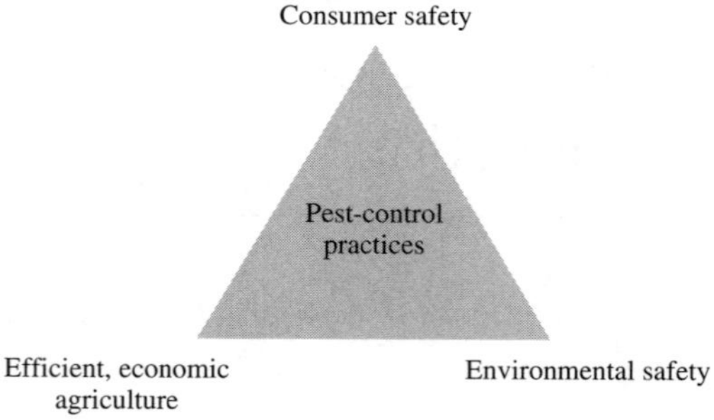

Figure 1.1 The underlying paradigm; the use of pest-control practices must be viewed as part of the impact of agricultural practices on the efficiency and economic viability of agriculture, environmental protection and human health

When considered singly, each of the approaches discussed in this book can generate an improvement in the effectiveness of pesticide use. Taken together, the combined improvements in chemistry, application technologies and the integration of chemical, physical and biological aspects with each other, one can genuinely envisage an optimisation in pesticide use without compromising the quality and efficiency of farming or consumer and environmental protection. As an introduction to this book, the broad subject has been split into three key areas: usage, application and formulation and complemented with the broader, 'multidisciplinary' approaches of RPU and complementary pest-control methods.

USAGE

Underpinning any attempts to change pesticide usage practices is the requirement to measure what happens now and to assess the effects of any change. The collection and interpretation of pesticide usage statistics provides the basis for monitoring the impact of changes in policy and legislation. However, the correct collection and interpretation of accurate usage information is complex, and requires both care and an encyclopaedic knowledge of local agricultural practices. Taking an example from United Kingdom usage data, Figure 1.2 shows the overall usage of a range of selected pesticides between 1983 and 1997.

A simple interpretation is that usage of this range of compounds has increased in the late 1980s and decreased in the mid-1990s. However, during the 15-year period many factors have contributed to the pattern observed. For example, a shift away from ethylene *bis*-dithiocarbamates towards fluazinam and propamicarb for blight control in potatoes, coupled with an overall move away from the use of maneb, and significant changes in the use of gamma-HCH (lindane) as a seed treatment on oilseed rape in the early 1990s. Such changes in agricultural practices, visible if the detailed information for each product is examined, are also compounded by changes in related agriculture regulation, such as the change

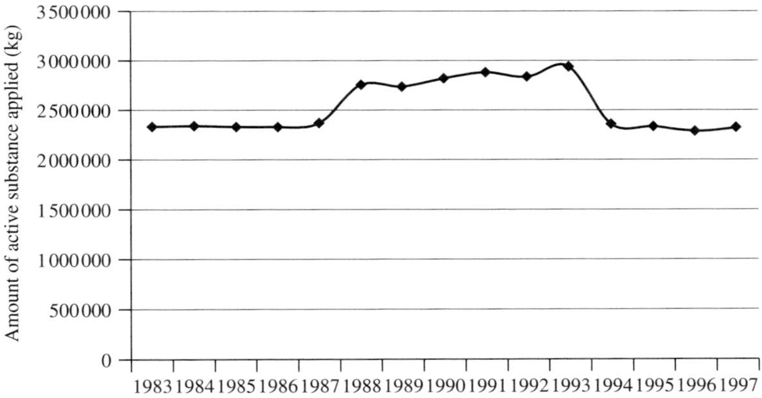

Figure 1.2 Usage of selected pesticides in the United Kingdom between 1983 and 1997. The data are expressed as kilogrammes of active substance applied. The compounds included are: carbaryl, dicofol, quinalphos, parathion, gamma-HCH, DDT, methoxychlor, amitrole, metribuzin, atrazine, linuron, diuron, 2,4,5-T, 2,4-D, nitrofen, trifluralin, fenarimol, sodium 2-phenylphenoxide, vinclozolin, ziram, zineb-ethylenethiuram disulfide zineb, thiram, maneb and mancozeb. Each has been alleged to exhibit some endocrine-disrupting properties

from 10% to 5% permissible set-aside in the UK in 1994. These subtle changes are all in addition to the overall effects of the changes in areas of crops grown, and changes in the use of such practices as tank-mixing and reduced dose application rates. The details are discussed in the next chapter. However, despite the complexity of factors contributing to changes in use, such survey information provides the best benchmark available by which to judge changes in policies and practices.

APPLICATION

Assuming that pesticides are used in agriculture, it is essential that the mechanisms by which chemicals are applied are as efficient as possible in targeting the active ingredient at the point at which it has the maximum effect on the intended objective, while simultaneously minimising exposure of the environment, bystanders and other non-targets. The body of this book seeks to explore the complex physical mechanisms and interactions that determine the effective, targeted application of pesticides across the agricultural spectrum. It also recognises that both efficacy and environmental protection are increased in direct proportion to the effectiveness of the communication of best practice to the end-user. This is sometimes overlooked.

Application technology does not begin and end at the point of use. It is recognised that control of handling chemicals prior to use, pouring and container design are essential elements in minimising operator exposure and point-source pollution arising from pre-use handling operations. Something as simple as the size of container neck and a degree of standardisation in design between manufacturers can have a significant benefit in handling pesticides. This theme continues with an examination of the technology and design of specialised spraying equipment and aerial application methods. The latter is often criticised as being imprecise and implicitly uncontrolled, but this not only has been shown to be untrue, but aerial application is a vital means of crop protection where farm sizes are vast. It also has a central use for human health programmes where land access would be hazardous and impractical, if not impossible.

FORMULATION

The challenges of application techniques are complemented by a consideration of pesticide formulation science. The development of a practical and stable formulation that has the precise chemical and physical properties to enable the delivery of the active ingredient to the site of action is central to this topic. Although the discussion in this book concentrates on the development of products for the application of pesticides to crops via the use of spray application equipment, it recognises that, firstly, the same considerations apply to other areas of application, and that, secondly, formulation science can and does influence the ultimate safety and efficacy of the product.

RATIONAL PESTICIDE USE AND COMPLEMENTARY METHODS OF CONTROL

Having developed a new active ingredient, formulated it in such a manner as to bring a stable product to the market with appropriate physical properties, and having access to well-engineered application tools, consideration must be given to when to employ a particular pest-control strategy. An in-depth discussion looks at the background to and current state of rational pesticide use (RPU), the role of integrated pest management (IPM) and developments in chemistries to maintain effective pest control whilst respecting safety and the environment. Tying these strategies together with communication with, and training of, farmers worldwide is quintessential to optimisation. There is also a link here back to usage statistics and crop disease forecasting, effectively ensuring that prophylactic pesticide use is minimised.

This theme is continued in the consideration of 'complementary pest control' methods. The term complementary pest control is used to describe two activities. Firstly, it can refer to those activities performed either immediately before or after a pesticide application, or to those performed at regular intervals in between treatments, with the objectives in both cases of increasing efficacy of control and of reducing the frequency of treatments. This in turn includes the integrated pest management (IPM) approach. Secondly, it can refer to alternative nonchemical treatments in their own right. The penultimate chapter in this book looks at both activities, and their relevance to modern pest control and agricultural practices. It considers the benefits of combined approaches and those which seek to eliminate the use of chemical agents completely.

CONCLUSION

In conclusion, strategies aimed at optimising pesticide use can be said to epitomise a pragmatic approach. In a simplified diagram of the pesticide regulatory approvals process (Figure 1.3), the subjects covered in this book cover the 'how, when and why' area. These considerations form an integral part of the approvals process and the post-approval, iterative reassessment of chemicals and practices.

- How should a particular chemical or practice be applied in such a way to minimise its impact on humans and the environment? What techniques and equipment are available to minimise impact?

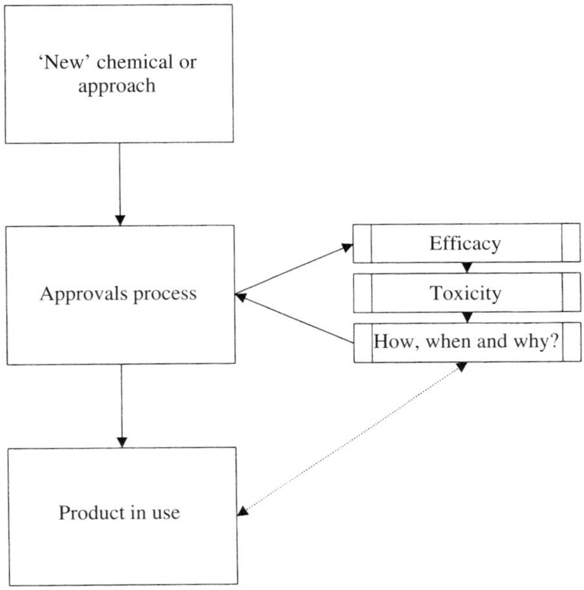

Figure 1.3 A simplified diagram of the approvals process. Details will vary from country to country, but in essence, any new or altered pest control agent will need to satisfy regulators that it works, it is not toxic to non-targets organisms (including humans) and the environment, and that there is a need for and safe means of using, the product or procedure

- When should a particular chemical or practice be applied and, more critically, when should it not be used because the benefits are outweighed by the risks? Are alternatives readily available?
- Why use a particular new chemical or practice when existing practices may be available and may be perfectly effective? Again, are alternatives readily available?

Finally, while alternative and complementary methods for pest control offer benefits under certain circumstances, they do not offer a panacea, and will not, in all probability, lead to a utopian, pesticide-free agriculture system. The best means of achieving optimisation in pesticide use and/or minimisation in chemical applications is the adoption of pragmatic approach based on a sound understanding of the chemistry, physics, biology and engineering involved in effective pest control.

2 Pesticide Usage Monitoring

M.R. THOMAS
Pesticide Usage Survey, Central Science Laboratory, Sand Hutton, York, YO41 1LZ, UK

The monitoring of pesticide use is a fundamental part of any policy on minimisation, from gauging the success of government initiatives, policy changes or fluctuations in support, through to environmental impact assessment at the field, farm, catchment, regional or national level. This chapter describes briefly the theory and methodologies behind usage monitoring, its value as a tool in minimisation, and some of the practical uses to which it may be put for measuring impact assessment at several different levels.

INTRODUCTION

During the last decade, there has been a growing global requirement for meaningful and accurate statistics on pesticide use. EUROSTAT, the statistical office of the European Communities, first published data on sales of pesticides in the environmental statistics yearbook for 1991, and further work was undertaken as part of the Dobris report (Stanners and Bourdeau, 1995) to produce a coordinated statistical appendix on pesticide use. Data, however, were poorly available, with specific information only obtainable for certain active substances or countries. It was also found that, in most cases, data were not very accurate, and different definitions of pesticides and their classification between countries made comparison difficult.

With the development of sustainable development indicators also moving into the role of pesticides and their impact on the environment, clearly, sound statistical information was required, particularly if the role of policy changes on pesticide use was to be assessed over time. Furthermore, an important target of the European Union's (EU) 5th Environmental Action Programme was the reduction of pesticide risk, but this would be impossible to monitor without sound information on changes in use over time.

Reductions in volume applied are meaningless with regard to risk, as many new active substances are applied at much lower rates per hectare than the older products that they are replacing, bringing about significant reductions in the weight

Optimising Pesticide Use Edited by M. Wilson
 ISBN: 0-471-49075-X

applied, without necessarily resulting in any reduction of use, as measured by area treated, or risk. From this point of view, the accumulation of sales statistics, and the general trends of reductions in weight used which they frequently show, can be seen to fall a long way short of providing the type of data required to allow meaningful assessment of the impact of policy changes on pesticide use and the consequences for the environment.

Few countries currently undertake regular monitoring of pesticide use through surveys of end-users. Within member countries of the Organisation for Economic Co-operation and Development (OECD), only the United Kingdom, the United States, The Netherlands and Sweden undertake regular surveys within the agricultural and horticultural areas. However, the importance of this work has been highlighted both within the EU, who commissioned the production of guidelines for the collection of pesticide usage statistics within member states, and the OECD, who accepted these guidelines and have since published them for use by all member countries (Thomas, 1999).

PESTICIDE USAGE MONITORING IN THE UK

As indicated above, the UK is one country with good pesticide use data. Official surveys of pesticide usage on a wide range of horticultural crops grown throughout England and Wales were begun in 1965, following concerns over the use of organochlorine insecticides on freshly consumed fruit and vegetables. By 1974, following the first survey of usage on arable crops, the first full programme of usage monitoring on all commercially grown commodities had been completed, and thereafter, surveys of individual crops were conducted on a variable three- to six-year cycle. Surveys of pesticide usage in Scotland also began in 1974, and were generally synchronised with those in England and Wales. These early cycles of surveys were summarised by Sly (1977, 1981, 1986) for England and Wales, and Cutler (1981) and Hosie and Bowen (1990) for Scotland. More recently, Thomas (1997) reviewed use throughout Great Britain over the period 1984 to 1994, while Thomas and Wardman (1999) reviewed use over the period 1986 to 1996. Surveys did not begin in Northern Ireland until 1989, but were synchronised from their outset to the timetable used in Great Britain.

Following the introduction of the Food and Environment Protection Act (FEPA) in 1985, post-registration monitoring of pesticide use became a legal requirement. Thus, in 1990 the government's independent Advisory Committee on Pesticides fixed the programme of surveys, such that arable crops were surveyed biennially, with surveys on all other commodities being undertaken on a rolling four-year cycle. From 1992, data were also presented for Great Britain, rather than reported separately, i.e. for Scotland, and for England and Wales. Surveys for Northern Ireland are still reported separately.

The pesticide usage survey (PUS) is a voluntary survey of a representative sample of agricultural and horticultural businesses located throughout the United

Kingdom. Each year, some 1500–2000 holdings are contacted and asked to supply details of current pesticide usage by crop or commodity category. The survey provides annual regional and national estimates of pesticide usage across all crops, although only certain commodities are surveyed in a given year.

The survey fulfils the legal obligation on the government under FEPA to monitor, following registration, the usage of pesticides. It is designed to obtain impartial accurate figures on the usage of all active substances, including areas of crop treated, weight applied, number of times treated, dose used, etc., and provides government and research with the most sophisticated and longest running data set of any country. Its methodology is respected throughout Europe and the OECD, and has been proposed as an EU and OECD standard (Thomas, 1999).

The survey divides all cropping into ten major commodity groups. Arable crops, representing around 86–90% of all pesticide use, are surveyed biennially, with the other nine commodity groups sampled every four years – two or three major surveys being undertaken each year. This effectively surveys all areas of pesticide use across agriculture and horticulture. The current timetable of surveys is given in Table 2.1.

Five broad methodologies of data collection are currently in use in different countries around the world. These comprise personal visits to a representative sample of farmers and growers to collect information on what they have used (e.g. UK, USA); telephone calls replacing personal visits (e.g. Sweden); postal surveys of a representative sample of farmers and growers (e.g. The Netherlands); collation of sales statistics (many countries), or compulsory returns from all users of pesticides (e.g. California).

Personal visits are used in the UK and have the advantage of accuracy, particularly where trained personnel are used, as the surveyor can go through all the potential uses which might have occurred, ensuring that the grower does not omit or forget anything important. The advantages and disadvantages of the other methodologies available are discussed in more detail by Thomas (1999).

The surveys use stratified statistical samples based on cropping and holding size. Survey samples are drawn from the Department for Environment, Food and Rural Affairs' (Defra) census data. For those crops where census data are not available, areas grown are supplemented by agronomic data from a variety of sources, including grower associations, levy bodies and other Defra data. A detailed description and methodology for sample selection is given in Thomas (2001).

Survey data are collected at the end of the cropping season for any particular survey, usually from October to March, using experienced pesticide usage surveyors. The information is collected on a field-by-field basis for each crop, including all aspects of the agronomy of the crop that may influence pesticide use in any particular survey. Where appropriate, this includes details of preplanting/drilling treatments, variety of crop, type of cultivation, seed treatment, dates of pesticide treatments, details of pesticide use including; name of product

Table 2.1 Survey cycle and frequency of surveying across 10 major crop commodities

Survey year	Arable crops	Grassland and fodder crops	Outdoor vegetables	Top fruit	Soft fruit	Hops	Glasshouse crops	Outdoor bulbs and flowers	Hardy nursery stock	Mushrooms
1996	P	–	–	S	–	S	–	–	–	–
1997	–	P	–	–	–	–	–	S	S	–
1998	P	–	–	–	S	–	–	–	–	–
1999	–	–	P	–	–	–	S	–	–	S
2000	P	–	–	S	–	S	–	–	–	–
2001	–	P	–	–	–	–	–	S	S	–
2002	P	–	–	–	S	–	–	–	–	–
2003	–	–	P	–	–	–	S	–	–	S

P = principal survey each year
S = secondary surveys

used, amount applied, volume of spray applied, use of adjuvants, and details of tank mixing.

The data collected are raised using information from the annual agricultural census returns to give national estimates of usage. More detailed methodology for each survey is published in the relevant Defra report, details of which are available on the World Wide Web at: www.csl.gov.uk/prodserv/cons/pesticide/intell/pusg.cfm.

A raising factor is generated for each cell (farm size within region), which is equal to the total area of farms within that cell divided by the total area of farms sampled within that cell:

$$Rf1_{\mathrm{sr}} = \frac{\text{total area of farms within size group s in region r}}{\text{total area of farms visited within size group s in region r}}$$

Any slight over- or undersampling of a particular crop within a region is corrected for, using a correction factor derived from the total area of that crop grown within the region divided by the raised estimate of crop grown in that region:

$$Rf2_{\mathrm{cr}} = \frac{\text{total area of crop c grown in region r}}{\sum_{1}^{n}(\text{area of crop c grown on farm n in size group s in region r} \times Rf1_{\mathrm{sr}})}$$

$RF2$ generally approximates to 1. Where a crop may not have been encountered in any particular region – an occurrence usually restricted to minor crops – a third raising factor is applied to correct for any such slight undersampling across regions, derived from the total area of that crop grown nationally divided by the raised national estimate of crop grown:

$$Rf3 = \frac{\text{total area of crop c grown}}{\left[\sum_{1}^{n}\text{area of crop c grown on farm n in size group s in region r} \times Rf1_{\mathrm{sr}} \times Rf2_{\mathrm{cr}}\right]}$$

THE ROLE OF USAGE STATISTICS IN MINIMISATION POLICIES AND IMPACT ASSESSMENT

The collection of a reliable set of usage statistics has value in many areas of research, legislation and agricultural support, and is far more than a simple statistical exercise in its own right. In Great Britain, usage statistics form an integral part of minimisation policy and the assessment of pesticide impact on consumers, operators and the environment. The most important of these areas are illustrated below and fall into nine main categories:

1 Provision of Annual Usage Statistics

In their simplest form, usage statistics provide information on national and regional levels of pesticide inputs to individual crops. Thus the total amount of any one pesticide used annually is available, together with the areas treated and the range of crops to which it has been applied. Table 2.2 gives an example of estimated use of the morpholine fungicide fenpropimorph, across all formulations, for Great Britain in 1999. Additionally, information on the total inputs of all pesticides to any one crop is also available. Table 2.3 gives an example of changes in pesticide use on carrots grown in Great Britain over the period 1986 to 1999 by pesticide group, area treated and weight applied. Both these may be broken down to provide a seasonal profile of use. Figure 2.1 illustrates the seasonal pattern of usage of major pesticide groups throughout a typical year (2000) on wheat grown in Great Britain, as dates of application are also collected. Such data are required at several levels:

- At a national level, to inform government of the current status of pesticide use. Following a number of recently reported 'pesticide scares' appearing in the press, concerning the carcinogenic, neurological or other undesirable effects of specific pesticides, it is vital that ministers have up-to-date information on their usage. This includes data on the product range in which they occur, the crops on which they are used, and the extent to which those crops are treated, ultimately yielding information on likely exposure of the population to the purported hazard. Without these data, the government could find itself embarrassed in being unable to defend the results of its own legislation. Data are also freely passed to universities, pressure groups such as Greenpeace, Friends of the Earth and the WorldWide Fund for Nature, and members of the general public.
- Within the EU, where EUROSTAT are trying to compile meaningful and comparative statistics across member states, partly in fulfilment of the EU's Fifth Environmental Action Programme, which set a target for the year 2000 of 'the significant reduction of pesticide use per unit of land under production...'. The success of this can only be monitored by collating reliable usage data over time.
- Within the OECD, where the Pesticide Forum, and in particular the Risk Reduction Group, have expressed a need for reliable usage statistics.
- Internationally, where FAO attempt to compile annual statistics across all countries under Article 1, para. 1 of the FAO Constitution, which stipulates that: 'the Organisation shall compile, analyse and disseminate information relating to nutrition, food and agriculture'.

Table 2.2 Fenpropimorph–estimated annual pesticide usage in Great Britain 1999 (all formulations). Data from Defra Pesticide Usage Surveys in Great Britain 1996/99

Survey	Crop	Treated area (ha)	Amount[1] applied (kg)	Area grown (ha)
Seed treatments				
Arable crops				
	Linseed	11 639	551	114 191
	Oilseed rape	247 578	1 600	505 423
	Total for survey	259 217	2 151	
Grassland and fodder crops				
	Kale, cabbage, rape, etc.	46	<1	18 806
	Total for survey	46	<1	
Average annual total–all seed treatments		259 263	2 151	
All other treatments				
Arable crops				
	Beans	2 332	606	110 590
	Oats	49 985	15 369	94 714
	Oilseed rape	1 594	120	505 423
	Rye	7 892	2 406	9 709
	Set-aside	316	17	310 604
	Spring barley	241 593	51 918	455 594
	Triticale	2 093	367	10 561
	Wheat	991 133	148 163	2 035 686
	Winter barley	594 621	126 694	760 497
	Total for survey	1 891 557	345 660	
Outdoor bulbs and flowers				
	Flowers for cutting	142	52	741
	Total for survey	142	52	
Grassland and fodder crops				
	Grass <5 years old	11 713	3 213	1 202 638
	Other crops for stockfeeding	2 223	553	14 524
	Total for survey	13 936	3 766	–
Glasshouse crops				
	Fruit (protected)	8	11	75
	HNS (protected)	2	1	254
	Total for survey	10	13	–

(*continued overleaf*)

Table 2.2 (*continued*)

Survey	Crop	Treated area (ha)	Amount[1] applied (kg)	Area grown (ha)
Hardy nursery stock				
	Fruit stock	1 733	705	776
	Mixed areas	25	17	2 010
	Ornamental trees	67	10	1 527
	Roses	274	200	860
	Shrubs, etc.	149	79	1 203
	Total for survey	2 247	1 010	
Hops				
	Hops	130	74	3 443
	Total for survey	130	74	
Soft fruit				
	Blackberry	7	1	194
	Blackcurrant –fresh market	77	48	244
	Blackcurrant –processing	503	286	1 124
	Gooseberry	162	93	261
	Hybridberry	2	1	229
	Raspberry	179	126	2 488
	Red/white currant	27	14	138
	Strawberry	2 225	1 554	3 887
	Total for survey	3 183	2 124	
Vegetables				
	Beetroot	173	124	1 722
	Brussels sprouts	346	196	5 576
	Carrot	6 948	5 053	11 771
	Herbs	34	25	1 690
	Leek	1 288	740	2 569
	Parsnip	1 524	1 143	3 360
	Total for survey	10 313	7 281	
Average annual total – all other treatments			1 921 518	359 980
Average annual total – all crops			2 180 781	362 131
Average annual total – all uses (kg a.s. applied)				362 131

[1]Weight of active substance (a.s.)

Table 2.3 Comparison of pesticide usage on Carrots in Great Britain (1986–1999): area treated (spray ha) and amount applied (t a.s.)

Chemical	1986		1991		1995		1999	
	ha	t	ha	t	ha	t	ha	t
Insecticides								
Carbamates	4 686	3.97	13 421	5.62	10 527	7.92	9 427	9.31
Organochlorines	0	–	0	–	0	–	2	< 0.01
Organophosphates	37 332	43.87	57 890	57.21	30 156	28.23	4 510	3.60
Pyrethroids	122	–	6 387	0.21	18 074	0.42	37 098	0.49
Other insecticides	0	–	0		37	–	1 098	0.17
Total – all insecticides	42 140	47.85	77 698	63.04	58 795	36.57	52 134	13.57
Fungicides	4 454	29.36	12 013	69.79	22 239	60.18	37 701	25.53
Sulfur	0	–	112	0.45	5 283	22.84	5 866	25.89
Herbicides	42 553	53.54	58 480	66.44	54 822	52.31	61 774	55.44
Molluscicides	2	<0.01	0	–	20	0.01	6	0.01
Total – all registered pesticides	89 148	130.75	148 304	199.72	141 159	171.91	157 482	120.44
Area grown	14 360	–	15 515		12 456	–	11 771	–

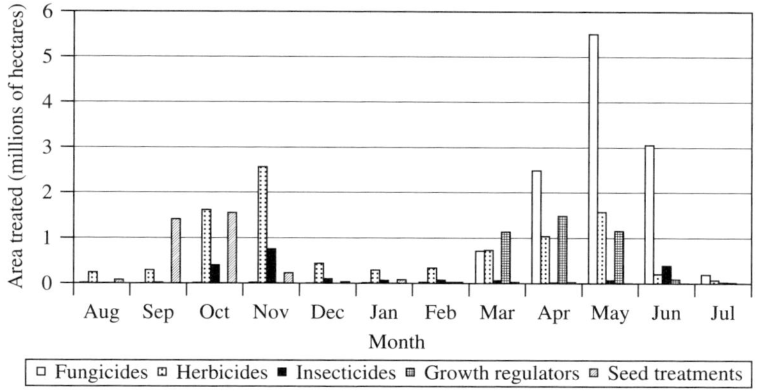

Figure 2.1 Seasonal profile of usage of pesticides on wheat, Great Britain, 2000 (active substance treated ha)

2 Providing Data-sets for the Development of Indicators of Environmental Impact

Usage data are critical for the development of indicators of the effects of pesticides on the environment, and data-sets over time are required in order to monitor the effects that policy changes may have on that impact. Both the former Department of the Environment, Transport and Regions (DETR) (Anon., 1996) and more recently the former MAFF (Anon., 2000), now Defra, have published standard sets of indicators on environmental quality which have utilised pesticide usage data. Programmes within the EU (Sectoral Infrastructure Projects in the Context of Environmental Indicators and Green Accounting) and OECD (Pesticide Forum: Pesticide Risk Reduction Project) are acutely aware of the necessity for sound usage data over time in order to fully develop such indicators.

3 Monitoring Changes Over Time

Once the collection of a regular set of usage statistics has been established, changes over time in use on particular crops (Figure 2.2), or of particular pesticides (Figure 2.3), can be monitored. These may result from several factors, some or all of which may interact to give annual variations in use:

- Annual differences in the weather, influencing the range of pest, disease and weed problems requiring control, or affecting the ability of the farmer to apply the pesticide under suitable conditions (Figure 2.4).
- The introduction of new molecules which may replace older, less-active pesticides, and may additionally be applied at much lower rates per hectare (Figure 2.5).
- Changes in the price of, or support level to, crops, thereby altering margins and making the use of pesticides more or less economic. The UK usage data-sets formed an integral part of the assessment of potential changes in CAP reform on predicted cropping levels and consequent pesticide use during the government's discussions prior to its implementation.

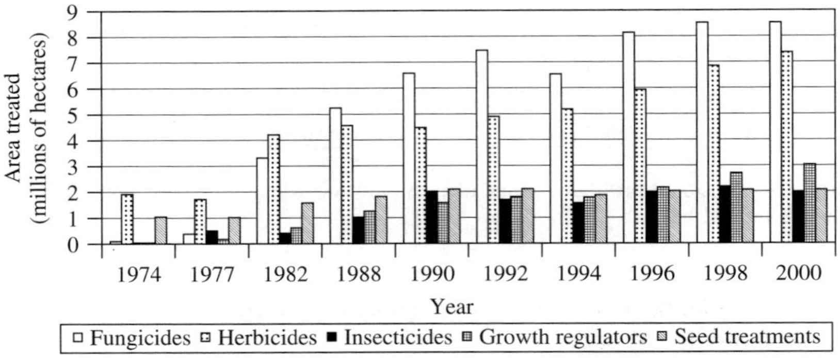

Figure 2.2 Changes in pesticide-treated area of wheat, Great Britain, 1974–2000 (formulation treated ha)

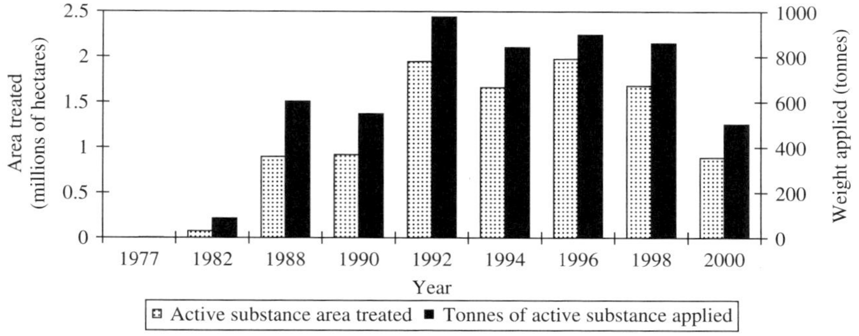

Figure 2.3 Changes in use of chlorothalonil on all crops, Great Britain, 1974–2000

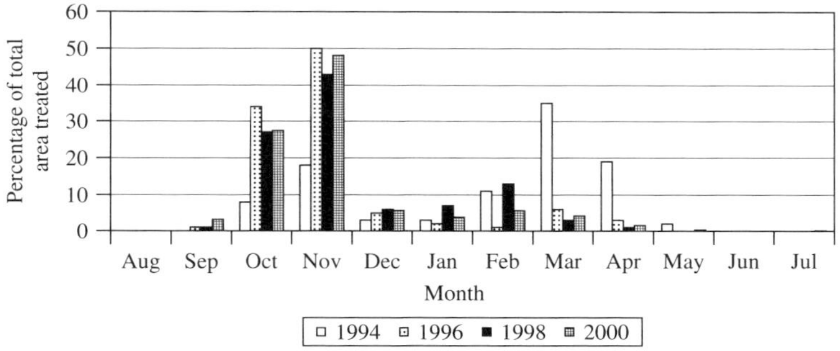

Figure 2.4 Annual variation in seasonal usage of isoproturon on wheat, Great Britain, 1994–2000

4 Providing Information as Part of the Review Process of Existing Pesticides

An essential part of the review process of a pesticide, currently under way for all existing pesticides within the EU, is a knowledge of the local and national uses and requirements for that pesticide (Orson and Thomas, 2001). If monitoring suggests that growers cannot compete without a particular pesticide, and no alternatives are available, this must be borne in mind during its review. Reliable usage data are fundamental to such appraisals and, as a suitable means of quantifying the effect of withdrawal, are regularly used during the review process in the UK. Alternatively, the demonstrated lack of use of a particular pesticide, coupled with the availability and uptake of safer or more benign alternatives, may hasten a pesticide's withdrawal. Furthermore, reliable estimates of the percentage of crop treated with a pesticide, the number of times that it is treated, and the actual rates of use, which may be significantly less than the maximum label recommended rates (Table 2.4), allow a more realistic assessment of consumption

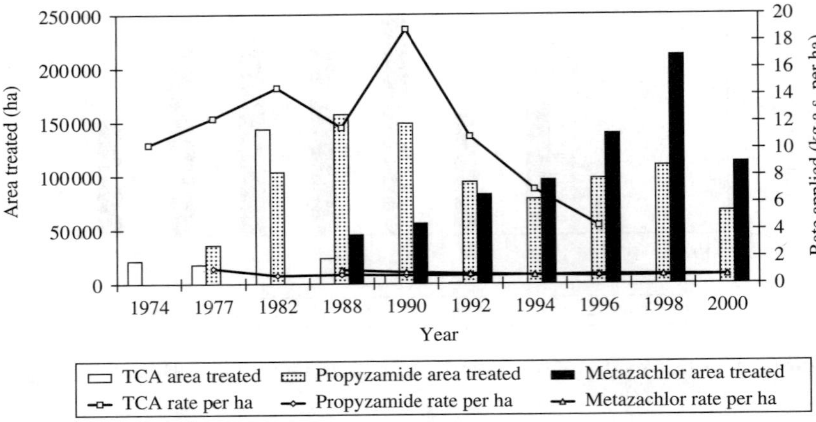

Figure 2.5 Temporal change in active substance spectrum for herbicide use in oilseed rape, Great Britain, 1974–2000

during consideration of acceptable daily intake. Without such data, the continued approval of products may be significantly affected.

5 Providing Information as Part of the Approvals Process of New Pesticides

During the approval of new active substances, usage data may provide a clear indication of the likely uptake of a new pesticide, knowing what pesticide(s) it is likely to replace and the current extent of its use.

Table 2.4 Relationship between actual area of Cox apples treated with carbendazim, amount applied and maximum amount permissible under label recommendations, Great Britain, 2000

Number of applications per season[a]	Area treated as percent of area grown	Average application rate[b] (kg a.s. ha/application)	Average total amount applied (kg a.s. season)	Percentage of maximum total amount allowed per season[c]
1	10.5	0.276	0.276	4.2
2	8.2	0.414	0.828	12.5
3	7.5	0.374	1.122	17.0
4	3.9	0.236	0.944	14.3
>4	1.7	0.090	0.900	13.6

[a] Label maximum number of applications per season = 12
[b] Label maximum application rate = 0.55 kg a.s. ha
[c] Label maximum total dose = 6.6 kg a.s. season

6 Refining Assessments of Predicted Operator Exposure to Pesticides

The establishment of a statistically sound data-set on usage allows an evaluation of likely operator exposure, as realistic work rates can be derived from the data collected, such as average field size, area sprayed per operator per day, amount of pesticide handled per day, etc. All of these factors are vital in refining predicted operator exposure models, and are discussed at length by Hamey (2001).

7 Monitoring the Potential Movement of Pesticides into Water

Data on pesticide usage can be used to assist in the monitoring of pesticide contamination in surface and groundwaters. For example, the EU aims to protect drinking water and groundwater through legislation, leading to widespread monitoring of pesticide residues in order to comply with these directives. Within Great Britain, usage data are used within a complex geographical information system developed by the Environment Agency (Williamson, 1997), containing maps of soil and groundwater, rivers and other waterways and water abstraction points. This is overlaid with current cropping and pesticide usage patterns, both geographically and seasonally, and, together with a database of pesticide properties and models of movement through different soils, is used to predict the likely appearance of pesticides at abstraction points to facilitate the monitoring of pesticides in water. By so doing, it is hoped to avoid unnecessary monitoring for pesticides which are unlikely to appear at a specific point or time within a given water body. It is important to note, however, that such methods can only be used to direct monitoring rather than substitute for it.

8 Monitoring Farmer Practice to Highlight Areas Where Use May Be Optimised

Data on farmers' actual use of pesticides may be examined to see if their current practices may be improved or optimised. For example, within Great Britain, the comprehensive database of farmer practice with regard to fungicide and insecticide use on winter wheat is being examined to identify where farmers may be using pesticide programmes inappropriately. This is being examined, particularly with regard to underutilising varietal resistance (Turner and Thomas, 1998) or inappropriately timed pesticide applications. Furthermore, there would appear to be some scope for reducing pesticide applications under certain circumstances. It is hoped that areas where clear savings can be made will be identified and targeted for further advice, in an effort to reduce inputs of pesticides to those crops. The technique should be applicable to many crops.

9 Providing Information For Residue Monitoring Programmes of Fresh Fruit, Vegetables, etc.

Usage data have provided the foundation for the development of residue monitoring suites for a wide range of home-grown produce within the UK to monitor compliance with Maximum Residue Levels (MRLs) and some of the implications of this are discussed further by Wilson and Thomas (2000).

- Where new monitoring programmes are being undertaken, usage data will illustrate the range of pesticides currently used on the crops to be monitored, and have allowed analytical suites to be tailored to consider only those pesticides likely to be encountered (Thomas, 1996).
- Where unusual or unexpected residues are found, usage data can confirm the results or invoke alternative methods to corroborate or invalidate the findings. For example, analysis of plums by HPLC with UV diode array detection indicated that 50% of samples contained residues of diflubenzuron, whereas usage data suggested that only 5% of the crop had been treated. These survey results prompted alternative analysis by LC–MS, which revealed that suspected residues were artefacts. In contrast, residues of chlorothalonil in lettuce, a non-approved use within the UK, were corroborated by survey data where such misuse had been encountered in the field. Internationally available survey data would allow countries to tailor their monitoring programmes for imported produce as well as home-grown foodstuffs.

CONCLUSION

Statistically derived estimates of pesticide use from surveys such as those outlined here and undertaken over the past 36 years, yield invaluable information for government, industry and research, and form an integral part of pesticide legislation in the UK.

The principal limiting factor to voluntary surveys such as these is the time taken to collect the data and the consequent burden on the farmer. Although more detailed information on sprayer technology, including cab type, boom width, handling mechanism, nozzle type, maintenance, etc., would be extremely useful, the survey's main role is the collection of usage information, and the time taken to collect this already approaches the grower's tolerance limit for voluntary surveys. Such additional information can, and has been, successfully collected by leaving a questionnaire with the farmer to fill in and return at his own convenience.

Nonetheless, although designed to produce estimates of total usage of pesticides, the statistically derived surveys undertaken in the UK do yield a wealth of information on pesticide use patterns across a wide range of holdings and crops. Not only do they provide a large, representative sample of actual usage patterns on individual holdings (see Hamey, 2001), they can also provide more general information on usage patterns by broad pesticide groups, individual pesticide classes or specific active substances (Thomas, 2001).

As other countries adopt their own usage surveys, the wider availability of usage statistics will not only facilitate international trade by providing more reliable information on the likelihood of residues in exportable produce, but also encourage the development of more reliable and broadly accepted indicators of environmental impact from pesticide use.

REFERENCES

Anon. (1996) *Indicators of Sustainable Development for the United Kingdom*. Department of the Environment, London: HMSO.

Anon. (2000) *Towards Sustainable Agriculture: a Pilot set of Indicators*. MAFF Reference Book PB 4583. London: MAFF Publications.

Cutler, J.R. (1981) *Pesticide Usage Survey Report 27–Review of Pesticide Usage in Agriculture, Horticulture and Animal Husbandry*, 1975–1979. Edinburgh: DAFS.

Hamey, P.Y. (2001) The need for appropriate use information to refine pesticide user exposure assessments. *Annals of Occupational Hygiene* **45**(1001), 569–579.

Hosie, G. and Bowen, H.M. (1990) *Pesticide Usage Survey Report 68–Review of Pesticide Usage in Agriculture, Horticulture, Animal Husbandry, Grain Storage, Forestry, Local Authorities and Water Authorities*, 1978–1986. Edinburgh: DAFS.

Orson, J.H. and Thomas, M.R. (2001) Impact of generic herbicides on current and future weed problems and crop management. *Proceedings of the Brighton Crop Protection Conference – Weeds*, 2001, **1**, 123–132.

Sly, J.M.A. (1977) *Pesticide Usage Survey Report 8–Review of Usage of Pesticides in Agriculture and Horticulture in England and Wales, 1965–1974*. London: MAFF. Publications.

Sly, J.M.A. (1981) *Pesticide Usage Survey Report 23–Review of Usage of Pesticides in Agriculture, Horticulture and Forestry in England and Wales, 1975–1979*. London: MAFF Publications.

Sly, J.M.A. (1986) *Pesticide Usage Survey Report 41–Review of Usage of Pesticides in Agriculture, Horticulture and Animal husbandry in England and Wales, 1980–1983*. London: MAFF Publications.

Stanners, D. and Bourdeau, P. (1995) *Europe's Environment – the Dobris Assessment*. European Environment Agency, Copenhagen.

Thomas, M.R. (1996) Pesticide usage surveys – towards a more efficient residue analysis. *Proceedings of the 1st European Pesticide Residue Workshop, Alkmaar, The Netherlands, June* (1996), pp 0–015, Inspectorate for Health Protection/Food Inspection Service, Alkmaar, The Netherlands.

Thomas, M.R. (1997) *Pesticide Usage Survey Report 100: Review of Usage of Pesticides in Agriculture and Horticulture throughout Great Britain 1984–1994*. MAFF Reference Book PB 2943. London: MAFF Publications.

Thomas, M.R. (1999) *Guidelines for the Collection of Pesticide Usage Statistics within Agriculture and Horticulture*. Paris: Organisation for Economic Co-operation and Development.

Thomas, M.R. (2001) Pesticide usage monitoring in the United Kingdom. *Annals of Occupational Hygiene* **45**(1001), S87–S93.

Thomas, M.R. and Wardman, O.L. (1999) *Pesticide Usage Survey Report 150: Review of Usage of Pesticides in Agriculture and Horticulture throughout Great Britain* 1986–1996. MAFF Reference Book PB 4188. London: MAFF Publications.

Turner, J.A. and Thomas, M.R. (1998) Analyses of fungicide optimisation potential in England and Wales through exploitation of wheat cultivar disease resistance. In: Proceedings of the 7th *International Congress of Plant Pathology, 9–16th August 1998 Edinburgh, Edinburgh, Scotland*, ICPP. pp. 4.9.19.

Williamson, A.R. (1997) POPPIE–A national scale system for prediction of pollution of surface and groundwaters by pesticides. *Pesticide Outlook* **8**(1), 17–20.
Wilson, M.F. and Thomas, M.R. (2000) Changes in the use of agricultural pesticides in the UK and their impact on the analysis of residues in food and the environment. Proceedings of the International Symposium on Crop Protection, 2000. *Mededelingen Faculteit Landbouwkundige and Toegepaste Biologische Wetenschappen* **65**(2b), 837–841.

3 Application Technologies

J.C. VAN DE ZANDE[1] **C.S. PARKIN**[2] **AND A.J. GILBERT**[3]
[1] *Institute of Agricultural and Environmental Engineering, IMAG, P.O. Box 43, 6700AA Wageningen, The Netherlands*
[2] *Silsoe Research Institute, Wrest Park, Silsoe, Bedford, MK45 4HS, UK*
[3] *Central Science Laboratory, Sand Hutton, York, YO41 1LZ, UK*

TARGETING PESTICIDE APPLICATIONS

The efficient application of pesticides and other xenobiotics to crops relies on targeting. The optimum use of pesticides requires not only correct timing, but also efficient transfer of active ingredients to those areas within a crop where the pests, weeds or diseases are located. Because a large majority of pesticide applications are made using liquid sprays, this chapter is devoted to the targeting of sprays where the whole field is treated. In a later chapter the selective application of pesticides to areas within a field, or 'patch spraying', is discussed.

Simple changes in spray application can result in dramatic changes to the distribution of pesticide within a crop providing significant changes in biological efficacy. However, the optimisation of application parameters is often neglected because the effects of changes are often complex and little understood by growers. However, in the face of environmental and economic pressures, field pesticide doses are being reduced. This will inevitably increase the pressure on growers to maximise the efficiency of their spray targeting, since the effects of poor application will be more evident (Enfält *et al.* 1997).

In order to understand the concept of targeting in field-scale applications, we will briefly examine the concept as applied to some typical scenarios.

ARABLE APPLICATIONS

Arable crops are normally considered as being plane or two-dimensional spray targets. Pesticide label recommendations for arable crops usually define doses and application methods, with little regard for the development of the crop. Limited application specifications such as 'apply 1 l/ha in 200 l/ha of water using a boom sprayer' are ubiquitous.

Optimising Pesticide Use Edited by M. Wilson
 ISBN: 0-471-49075-X

However, from the beginnings of spray research, it was understood that the quantity of pesticide reaching the canopy of arable crops could be influenced by application method. Courshee (1960) recorded that, although only 20% of the volume sprayed by a typical application was deposited on the crop canopy, it could be influenced by crop growth and the volume applied. With the higher volume rates applied at that time, runoff spraying was very common, and losses to the ground underneath the crop were very high. Courshee suggested that, proportionately, spray deposition on the crop increased with decreasing spray volume.

Clearly, through the growing season, crops change considerably in their size and structure. A cereal crop, which may have provided little ground cover and flat spray targets early in the season, rapidly develops into a dense canopy with significant structure as it matures. To ignore this structure is to neglect the nature of the crop and miss an opportunity for optimisation. The situation is somewhat different in row crops such as potatoes and beet. Row crops have been more easily identified as being three-dimensional targets, and for many years, growers have accepted the need to direct their sprays towards the intended target and to account for growing foliage.

TREE AND BUSH CROP APPLICATIONS

Bush crops such as vines, and top fruit, provide linear vertical targets for spray and require particular consideration. Specialist equipment has been developed and there is awareness amongst growers of the effect of foliage density. However, since many bush-crop sprayers use air-jets to distribute the spray, improved targeting requires a practical knowledge of the interaction of air-jets and foliage.

Full-size trees, such as are found in traditional orchards, provide a complex and three-dimensional target. Spray deposition often relies on air assistance, and the influence of the structure of the target is great. Thus, with applications to trees, the prospect for improving the targeting sprays and the likelihood of poor application is great.

MAXIMISING DEPOSITION ON TO TARGET SITES

The aim of an optimised pesticide application must be to maximise the proportion of the active ingredient of the spray that deposits on the target site. The target site must be that area within the crop where the pesticide is most biologically active. It can be seen that the 'target' not only varies with the pest[1] but also with the pesticide. Since different pesticides have different modes of action, the choice of spray target will also vary with the pesticide. An insect pest may be controlled with a direct-acting contact material and the target will be the pest itself. A highly translocated pesticide that acts as an insect stomach poison will require a more indirect route and require another choice of target. Weeds may be

[1]Pest in this context means insect pest, disease or weed.

controlled by a pre-emergence application of a residual herbicide to bare soil, or may be controlled by a post-emergence spray by direct contact with the growing plant. In each case, consideration must be given to both the pest and the chemical. For an optimum application, this should then affect the choice and operation of the sprayer. Using the terminology of Joyce *et al.* (1977), the optimum use of pesticides requires a target-orientated application.

FUNDAMENTALS

In order to optimise pesticide application, knowledge of its performance is required. We will examine here some of the fundamental features of pesticide application techniques, so that later we can introduce a systematic approach to targeting and examine how specific applications in specific crops can be targeted. More detail on pesticide application fundamentals is found in Mathews (1992) and Mathews and Hislop (1993).

DROP PRODUCTION

Sprays for crop protection are generated by various means (Lefebvre, 1993), but most sprays used in agriculture are generated using hydraulic pressure. Flat-fan and hollow-cone nozzles are the most common, but deflector or anvil nozzles are used. Hydraulic flat-fan nozzles are available in a range of sizes, producing sprays varying from Very Fine to Coarse (Southcombe *et al.*, 1997). Typical drop-size distributions are invariably wide, and drop size distributions varying from 10 to 1000 μm diameter are not uncommon.

Although wide drop-size distributions can sometimes be an advantage (Hislop, 1983) the large numbers of small drops produced in hydraulic nozzle sprays can result in spray drift and inadequate targeting (Miller, 1993). Rotary atomisers (Bals, 1975) can reduce the breadth of drop-size distributions and provide a more targeted size distribution. They have not been widely adopted in broad-acre ground crops, because their spray volumes and drop trajectories often cause control difficulties, but they have found some acceptance in orchard sprayers.

Advanced nozzle designs, such as twin-fluid, pre-orifice and air-inclusion nozzles, can also be used. Most are designed to reduce spray drift. Atomisation in twin-fluid nozzles occurs because the interaction of air and liquid. Different spray qualities can be produced by changing both liquid and air pressures. Low spray volume rates (75–150 l/ha) and high work rates (ha/hour) are possible.

DROP BEHAVIOUR

Drops in a typical hydraulic nozzle are formed in a liquid sheet that travels at 15–25 m/s. Following sheet disintegration, drops move in an air-jet caused by the interaction of the spray plume and the surrounding air. Close to the nozzle, all drops move at the same speed, but, as the air-jet decays, fine drops with their greater drag to mass ratio become detrained. They can then become influenced by atmospheric air movements and cause spray drift. Lower spray volumes usually require smaller orifice nozzles that, in turn, produce finer sprays and increase the potential for spray drift.

Many modern sprayers are equipped with additional features designed to decrease spray-drift and improve deposition. Features include air assistance, where airflow, created by a fan placed on the sprayer, creates an air curtain behind the spray. The air is usually ducted via a flexible sleeve mounted on the boom. Fine drops leaving the nozzles are entrained in the air curtain and projected towards the crop (Miller and Hobson, 1991).

Large drops, with their greater kinetic energy, can cause problems as they impact on crops. Drops greater than 200 μm diameter (Brunskill, 1956) have the potential to cause spray runoff and contamination of the soil. With the lowering of spray volume rates, and the use of more fine nozzles, this problem has reduced in importance in recent years, but the trend towards the use of coarse sprays for drift reduction may reintroduce the problem in the near future.

CROP GROWTH

As crops develop from seeds to maturity, their development can be described by the growth stages that can be distinguished during the growing season, and there are several established scales in use. Crop growth stages are defined by the number of developed leaves (e.g. sugar beet), the number of branches (e.g. potatoes and cotton), or by a combination of stem elongation, number of developed leaves and reproduction organs (e.g. cereals). Decimal crop growth stage codes have been defined to describe the development of potatoes (Hack *et al.*, 1993) and cereals (Zadoks *et al.*, 1974).

During the growing season, as the crop develops, leaf mass increases until fruit-setting or tuber-filling, when leaf mass normally starts to decline. At this point, translocation occurs from leaf to reproduction organs – the product to be harvested. This means that the amount of leaf mass, and therefore the area to be covered by the pesticide, alters during the growing season (Figure 3.1). These changes also have an effect on where, for example, disease occurs. In dense crop canopies, fungi and spore survival is higher than in an open-crop canopy. Since the leaf area of an arable crop can cover five times the soil surface area (Leaf Area Index, LAI = 5), the spray volume may be required to cover not just one hectare per hectare of land, but five hectares per hectare of land. Because the target sites of diseases are not often known, the spray may be required to cover the whole of the plant tissue, on both sides of the leaves, requiring as much as 10 ha coverage per 1 ha of land. Yet, despite these large areas of plant tissue, shielding of the soil by leaves is never complete. Drops can penetrate through dense crop canopies and reach the soil. Losses to soil depend on spray quality and the application system, in relation to crop type and its development.

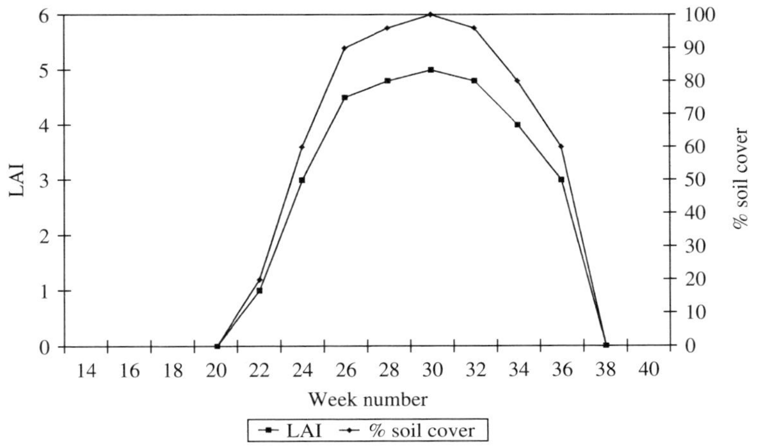

Figure 3.1 Crop growth during growing season, leaf area development and associated soil coverage (after Beukema and van der Zaag, 1979)

BIOLOGICAL TARGET SITES

Any pesticide must reach its site of intended biological action to be useful. This is ultimately the target for the pesticide, because if it does not reach this site it cannot work. The range of potential pest species, compounded by variations in the possible sites of pesticide action among these species, creates an almost overwhelming complexity that could prevent any meaningful general description of the location of target sites. However, sites of biological action are predominantly located within the pest organism, which will require the pesticide to be ingested (e.g. into the gut) or further absorbed (e.g. into nervous tissue). The pest surface is often regarded as a biological target site, either when the surface is the site of biological action or when the pesticide is absorbed through the surface to act internally.

To describe the main characteristics of biological target sites and alternative targeting pathways, it is useful to discuss groups of pests that have similar characteristics. Perhaps the most obvious distinction lies between pests that present either mobile or static targets for pesticides. The life stage of the species to consider is that which is vulnerable to pesticide action. Mobility creates both challenges and opportunities for targeting. Although moving targets may be harder to intercept if their surface itself is the target, this motion can be helpful should a sufficient dose of pesticide need to be accumulated by the target pest (e.g. found and eaten). Pesticide must reach static pests directly or indirectly, depending upon the availability of supplementary environmental forces that can redistribute residues toward their target. There are some classic scenarios that illustrate these distinctions. Insecticides may be placed as baits to be orally ingested by mobile targets (e.g. locusts). Herbicides and fungicides that act upon germinating seeds or spores may be placed upon the surface where the weeds or fungi will try to develop (e.g. soil or crop foliage, respectively). Translocated pesticides (of all types) may be applied to a suitable matrix within the translocation system responsible for moving the pesticide towards its targets. Foliar-applied pesticides need to be deposited upon the foliage in question, which, for nonselective herbicides, demands a selective application technique to exploit differences in foliar morphology (e.g. tall thin grassy stems vs. shorter broadleaved plants).

The form and distribution pattern required of deposits must facilitate the pesticide reaching its intended biological target, although an additional consideration is the need to allow for protection of non-target species, possibly using physical means of selectivity. In general, mobile pests will require less evenly distributed deposits, because they can intercept them, especially if some means for adding an attractant to the bait is available. Prophylactic deposits (e.g. pre-emergence herbicides) have to be very well distributed in order not to leave gaps where control will not be effective. Static pests may require an adequate dose to be deposited on their surface, but the exact location of that deposit may be more or less critical, depending on the way that the pesticide acts or is absorbed.

A SYSTEMATIC APPROACH TO TARGETING

Agricultural pesticide application practice covers treatments ranging across the whole crop growth cycle, from protecting seeds pre- and post-planting, through safeguarding growing plants, to protecting the yield both before and after harvest. Products may be applied in the widest possible variety of physical forms. Those applied directly as formulated may involve solids, liquids and gases/vapours, while those applied following preparation of use, as diluted mixtures, generally involve liquids (usually aqueous). The propensity for the applied formulation or mixture to be dispersed, in order to achieve appropriate distribution, will vary with the physical form that is applied. The intrinsic mobility of solids is less than that of liquids, which are, in turn, less mobile than gases. It should be noted that more intrinsically mobile forms offer a greater capability of reaching challenging targets, but are correspondingly more difficult to manage in terms of hazard containment.

The scope to alter both the efficiency and safety of application through control of the fate of applied material has been explored in a classification of application equipment by hazard (Parkin *et al.*, 1994). Application techniques as well as equipment are required to be taken into account for such a classification, which covers all four possible distinct styles of application i.e. *direct, liquid, solid* and *space*. The 'direct' style of application, which is typified by the need to bring treated items into intimate contact with pesticide, is exemplified by admixture processes, e.g. seed treatments. Direct applications take place under highly controllable conditions and often within relatively confined environments, which can potentially mitigate the exposure risks from applied material (presented either in

Table 3.1 Application equipment classification (after Parkin *et al.*, 1994)

Aspect	Possible descriptors
1. *Platform*	Ground based/aerial/water-borne
2. *Carriage*	Fixed installations/vehicle mounted/pedestrian carried
3. *Direction*	Downwards (placement)/upwards, outwards (broadcast)
4. *Distribution*	Types of atomiser for sprays/types of spreader for solids
5. *Form*	Qualities of liquid spray that might be used (e.g. BCPC Fine, Medium or Coarse)/corresponding qualities of solid granule dust or powder
6. *Assistance*	Shrouding or ducting by either mechanical means or by air-assistance; use of electrostatic charging.

liquid or solid form). The 'liquid' style of application is typified by the use of a liquid vehicle to carry the product, usually as a spray, to cover treated surfaces with deposits varying in their size and frequency. The 'solid' style of application also achieves distribution of product in more or less finely divided forms, in a similar way. In the liquid and solid application styles the application equipment (i.e. what is used) and technique (i.e. how it is used) determine the fate and behaviour of the applied product. The 'space' treatment style of application is self-explanatory – utilising the gaseous or vapour phase, or particles of neutral buoyancy, to transport applied product throughout an enclosed space.

Despite the inherent complexity of such a wide-ranging classification task, only six distinct general descriptive aspects are needed to cover the entire range of possible application systems and techniques that apply to all four styles of application. These are shown in Table 3.1.

By selecting suitable descriptions for each aspect, an objective description of all feasible application methods (equipment and technique) is possible. This equipment and technique classification scheme outlined in Table 3.1 offers a taxonomy that would allow objective differentiation to be made between seemingly similar systems, based upon their measured performance judged against other comparable systems.

EXAMPLES OF TARGETED APPLICATIONS

ARABLE CROPS

The most commonly used application systems in arable crops are boom sprayers equipped with equally spaced hydraulic nozzles (Robinson, 1993). The quality of the spray influences the quantity deposited on the crop and spray-drift. The combination of forward speed, nozzle type and pressure defines the sprayed volume. Volume rates have been reducing for several decades. While in the 1950s it was common to spray 1000 l/ha, nowadays 100–300 l/ha is more common.

The amount of the spray liquid deposited on to the soil, or on to crop leaf surfaces, can be expressed as a percentage of the applied dose (or spray volume) per unit surface area. In most cases this is measured by quantifying the amount of chemical tracer (often fluorescent) eluted from collectors placed in the crop canopy or directly from leaf tissue (Cooke and Hislop, 1993). Although many references report on spray deposition, the primary aim of much of this work is not to quantify spray distribution but to compare spraying techniques. Many researchers present spray deposition data only in order to assist biological comparisons. Some work quantifies the distribution of pesticide within the plant canopy (Figure 3.2), with deposits measured at different leaf levels (e.g. top, middle, and lower) and different leaf surfaces (i.e. upper, lower). Since leaf surface area is greater than the surface on which the crop grows, the leaf area index (LAI) is then used to produce a mass balance for the spray emitted. Total deposition on the crop is the sum of the deposited spray of both leaf sides times the LAI per leaf layer.

Cereals

The growth stages of cereals (Zadoks *et al.*, 1974) can be used to classify spray deposition during the growing season. The spray deposited on a cereal crop (Cooke *et al.*, 1985; Zande *et al.*, 1994b), and the soil surface beneath, are tabulated with respect to growth stage in Tables 3.2 and 3.3. In Figures 3.3 and 3.4 the effect of two spray volumes and the use of air assistance are presented for both soil and plant deposition. Spray deposition on the ground below cereals was on average 52% for the whole growing season.

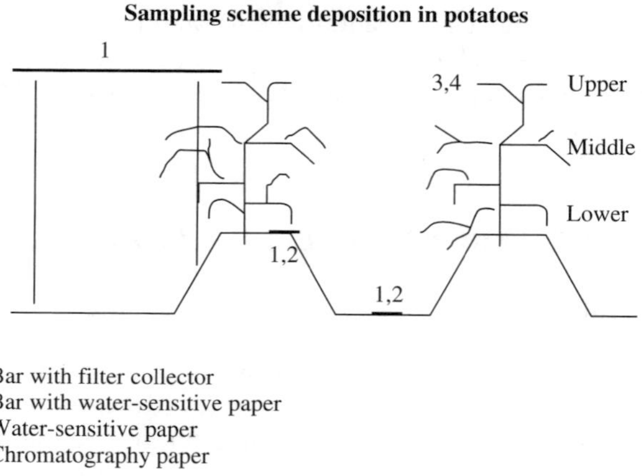

Figure 3.2 Schematic representation of deposition measurements (after van de Zande *et al.*, 1994a)

Table 3.2 Spray deposition (% dose) on the total plant canopy for cereals (wheat and barley) for different growth stages (Zadoks *et al.*, 1974) and different spray techniques[a]

Growth stage	22–29	30–32	33–37	38–45	49–59	61–69	Average
Conventional	45	61	84	86	63	64	67
Air assistance	57	64	84	92	87	61	74

[a] Spray volume measured in litres/ha.

Table 3.3 Spray deposition (% dose) on the soil surface underneath a cereal crop (wheat or barley) for different growth stages (Zadoks *et al.*, 1974 and different spray techniques[a]

Growth stage	14,22	21–32	31,32	32,37	45	49–61	Average
Conventional spray technique	95	78	28	–	8	15	52
Air-assistance spray technique	–	84	13	–	–	4	48

[a] Spray volume measured in litres/ha.

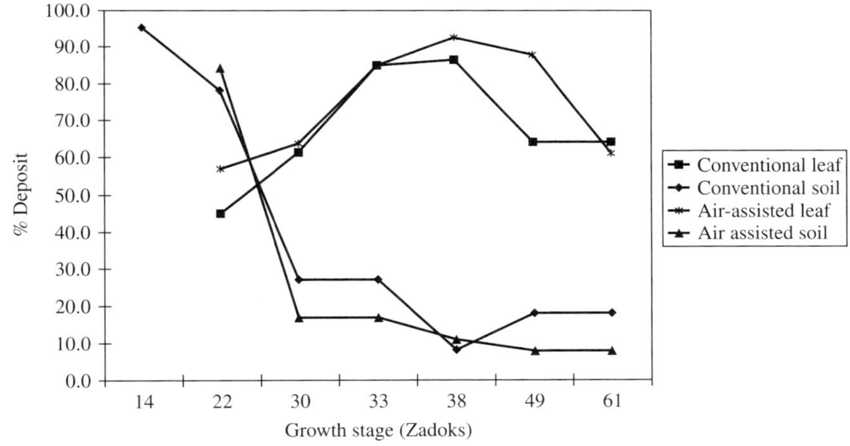

Figure 3.3 Total spray deposit (%) on cereals and soil underneath for conventional and air assisted spraying at different growth stages (Zadoks *et al.*, 1974)

At a specific growth stage it appears that there are also differences in spray deposition because of spray volume (Zande *et al.*, 1998). Growth stages 22–29 and 38–45, and low-volume sprays (50 and 100 l/ha) gave higher deposition on plant surfaces than higher volumes (200 and 400 l/ha). However, at growth stages 49–59 and 61–69, higher spray volumes (150, 200 and 300 l/ha) appeared to have greater deposition on leaf surfaces than lower spray volumes (50 and 100 l/ha). This effect is presented in Figure 3.6 for the spray volumes 100 and 200 l/ha.

Maximum soil deposit in cereals occurs early in crop development. More than 50% ground deposition is found up to growth stage 21–32 (Taylor and Andersen,

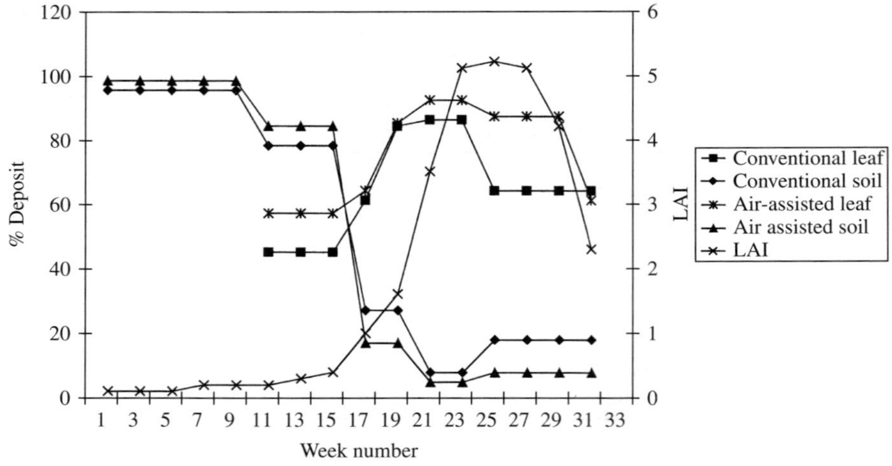

Figure 3.4 Spray deposition on leaf canopy and on soil surface related to Leaf Area Index when spraying a winter wheat crop

1991). Later, the crop develops rapidly, and ground deposits can decrease to 28% (Cooke *et al.*, 1990; Ringel, 1991, Van de Zande, 1995). At growth stages above 45, soil deposits further decrease to 9%, but a small increase to 15% occurs at growth stages 49–61. The use of air-assistance decreases soil deposition at growth stage 31 from 50% to 13%, and at growth stage 61 from 15% to 4%.

Spray deposition on leaf canopy in cereals is on average 67%. Increased plant deposition occurs with the use of air-assistance in almost all growth stages (Figure 3.3). Air-assistance gives an average of 74% deposition throughout the growing season, this being 7% more than with conventional spraying.

Soil and leaf deposition in winter wheat varies throughout the period from early January, when the crop is first sprayed, until harvest around August. Initially, when crop structure is very open, almost all the spray is deposited on the ground. As crop development begins, as is indicated with a theoretical LAI development based on the WOFOST crop growth model (van Diepen *et al.*, 1989), soil deposition diminishes and leaf deposition increases. In the period from week 11 till week 17, there appears to be an overestimate of the leaf deposition. This may be because, in the referenced literature, data are presented as total deposit independent of LAI at the time of spraying.

Potatoes

During the growing season of potatoes, three distinct stages in crop growth can be distinguished:

- *A* where the leaf canopy is in distinct rows;

- *B* where the leaf canopy covers soil surface completely;
- *C* where the leaf canopy is decreasing because of ageing;

During each of these periods, there are changes in coverage of the soil by the crop and LAI. This affects spray deposition both on the leaf canopy and on the soil beneath. This is shown in Figures 3.6 and 3.7 for conventional hydraulic spraying and air-assisted spraying. In Figure 3.5 the results are presented for the spray deposition in the potato plant.

In growth stage *A*, LAI is between 1 and 2 (Zande *et al.*, 1998) and soil coverage is 20–50%. Deposition in the crop leaf canopy in growth stage *A* is on average 46% (Zande *et al.*, 1998). In growth stage *B*, where soil coverage is complete and the crop is growing vigorously, LAI rises up to 5.1. Average deposition of spray in crop canopy at this stage *B* is 68% (Heer *et al.*, 1985; Porskamp *et al.*, 1993a; Wretblad 1997; Zande *et al.*, 1994a, 1998, 1999, 2000).

Later in the growing season, when plant stems lie between the ridges (stage *C*) LAI falls again to 1–2. Deposition on the potato plant is also reduced and, for conventional spraying, it is around 56% (Porskamp *et al.*, 1993a; Wretblad, 1997; Zande *et al.*, 1994a, 1998, 1999, 2000).

Using air-assistance on field sprayers changes the deposition pattern within the potato crop canopy. Penetration of the spray into the canopy is increased and by using air-assistance more spray is deposited at the middle and lower levels (Figure 3.5). At early and late growth stages (*A* and *C*) deposition in the crop using air-assistance is about 6–10% lower than with conventional techniques. With a fully mature canopy (*B*), the use of air assistance increases spray

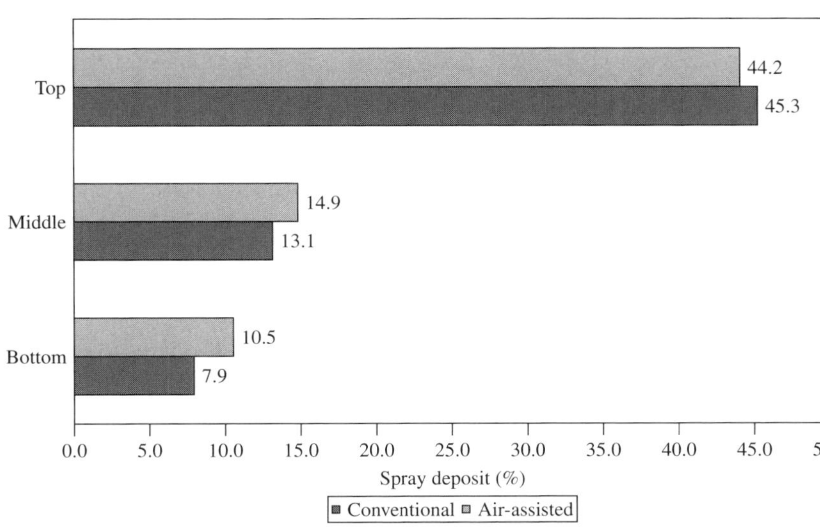

Figure 3.5 Effect of spray technique (conventional, air-assisted) on spray distribution over different leaf layers (top, middle, bottom) within the potato leaf canopy

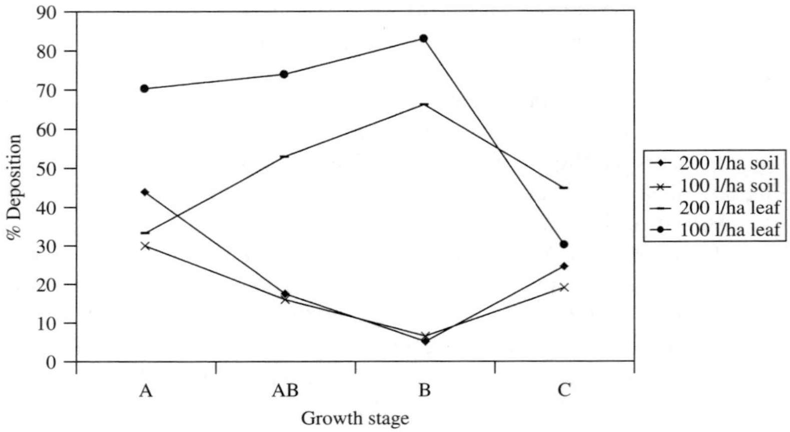

Figure 3.6 Spray deposition (% dose) on the total potato plant and soil surface underneath a potato crop for different growth stages, and different spray volumes

deposition in the crop by an average of 4%. Improvements in deposition with air-assistance are especially good with fine sprays and low volumes (100 l/ha). The effect of spray volume is presented in Figure 3.6.

During the growing season of the potatoes, as leaf coverage changes and spray deposit on the crop varies accordingly (Table 3.5) a change in deposition on the ground also occurs. In Table 3.4 the spray deposit is presented on soil surface underneath the potato crop. At growth stage *A*, deposition on the soil surface between the potato rows is at full-dose (100%). Averaged with the deposit underneath the plant rows on top of the ridges, deposition of the spray is still 39% (Zande *et al.*, 1998). On a completely covered soil surface in stage *B*, spray deposition on the ground decreases to 7% (Cooke *et al.*, 1990; Porskamp *et al.*, 1993*a*; Zande *et al.*, 1998).

At growth stage *C*, average soil deposition is 23% (Zande *et al.*, 1998). The use of air assistance increases spray deposition on the soil surface in all growth stages (Figure 3.7) resulting in an all-season average increase in soil deposition from 22 to 27%.

Table 3.4 Spray deposition (% dose) on potato plant for different growth stages, and different spray techniques[a]

Growth stage	A	AB	B	C	Average all season
Conventional spray technique (average)	46	60	68	56	57
Air-assistance spray technique (average)	40	60	72	46	54

[a] Spray volume is measured in litres/ha.

Table 3.5 Spray deposition (% dose) on soil surface underneath a potato plant for different growth stages, and different spray techniques[a]

Growth stage	A	AB	B	C	Average all season
Conventional spray technique (average)	39	17	7	23	22
Air-assistance spray technique (average)	50	21	10	27	27

[a] Spray volume is measured in litres/ha.

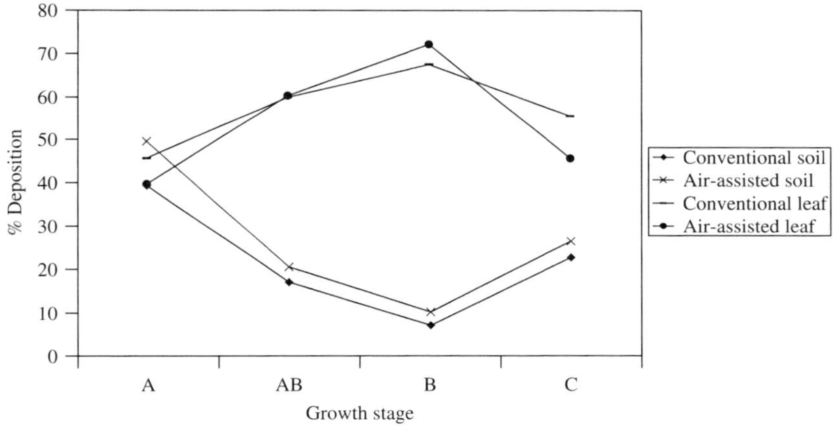

Figure 3.7 Spray deposition (% dose) on the total potato plant and the ground underneath for different growth stages, effect of air assistance

It is expected that at the start of the season more spray will be deposited on the soil surface than on the crop. Since most of the spray deposits on the crop when the maximum LAI occurs, and this diminishes towards the end of the season, we can expect that deposition on the soil will show the inverse of this. From the data found in the literature, these effects are shown in Figure 3.8, for potatoes (Zande *et al.*, 1998).

The data found in the literature can be categorised according to the time of the season and growth stage. Results in Figure 3.9 show a time-sliced distribution of spray liquid on crop canopy for sprays applied for late blight control in Europe between June and September. As LAI development during the growing season is not defined in the referenced literature, a simulated development of LAI was taken from a crop growth model WOFOST (van Diepen *et al.*, 1989).

Sugar Beet

May (1991) describes an application in a sugar beet crop, where 30% of the soil-surface was covered with the sugar beet crop. At this growth stage, 11–35% of the

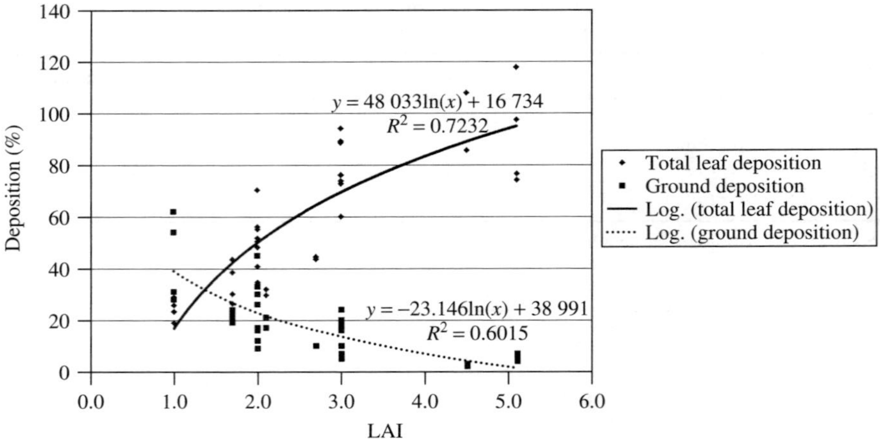

Figure 3.8 Spray deposition on potato leaves and on the ground as percentage of the dosage, related to LAI of the crop

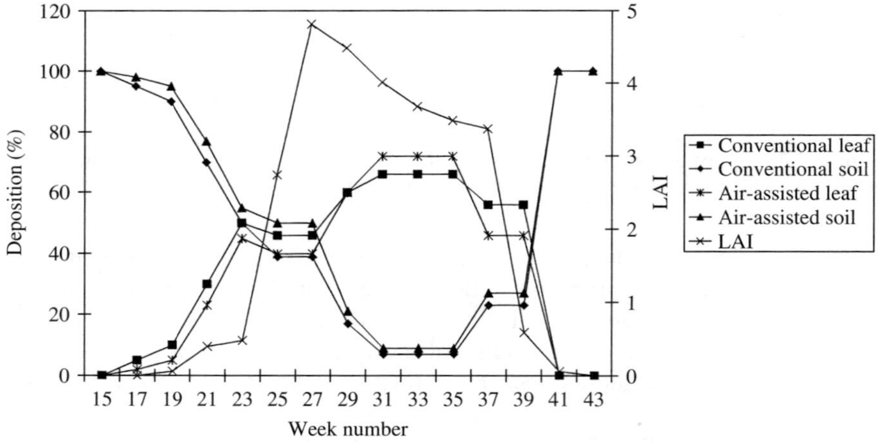

Figure 3.9 Spray deposition on crop canopy and underneath on soil surface, related to Leaf Area Index when spraying potatoes (after Zande *et al.*, 1998)

applied spray volume was recovered on the sugar beet plants with conventional spraying (90 l/ha). Soil deposition underneath the crop ranged from 34 to 51%. The use of air-assistance, on half power, gave similar amounts of retention of the spray on the crop and on the ground. Full-power air-assistance decreased spray deposition on the crop, but also increased soil deposition.

Brussels Sprouts

Heer *et al.* (1985) measured spray deposition on the buttons of a Brussels sprouts crop. On average, 2.3% of the sprayed volume (600 l/ha) was deposited on the buttons. The amounts deposited on the leaves and the stems were not investigated. Total plant deposition was therefore not quantified. Cooke *et al.* (1990) found soil deposition underneath a Brussels sprout crop to be 13%. The use of air-assistance did not change ground deposition, possibly because a finer spray had been used with air-assistance than with the conventional spraying. Plant deposit was expressed as ng/g dry weight per button on the different heights on the stem. Translation of the data into percentage of applied dose cannot be made, because of insufficient data.

Onions

The spraying of fungicides on to onions at 50% soil coverage and LAI = 3.8–4.1 resulted in an average of 33% deposition on onion leaves (Zande *et al.*, 1996). The use of air-assistance on the sprayer increased by 33% spray deposition on the leaves. On average, soil deposition was 40%. An increase in soil deposition by the use of air-assistance occurred only with the 200 l/ha volume rate, and not with the 100 l/ha spray volume. It should be noted that difference in spray volumes was achieved by a change in driving speed, and not the more usual change in nozzle sizes.

Flower Bulbs

Early in the season, spray deposition on lily leaf tissue (Table 3.6) is higher using conventional sprayers than with air-assistance (IJzendoorn *et al.*, 1998). Later in the growing season, there was little difference between the two spray techniques. Applying pesticides diluted in 150 or 300 l/ha instead of in 600 l/ha increases spray deposition on lilies by a factor of two-thirds (total plant deposition was respectively 57, 50 and 31%).

Porskamp *et al.* (1997) found, when spraying lilies in October, that average soil deposition (Table 3.7) was 61% with a conventional sprayer, whereas, with

Table 3.6 Spray deposition (% dose) on the leaf canopy of a lily crop (bed-grown) for different periods during the growing season, and for different spray techniques[a]

Lily growing season	May	June	(July) August	(Sept.) October	Average
Conventional spray technique	93	35	46	61	59
Air-assistance spray technique	70	31	32	–	44

[a] Spray volume measured in litres/ha.

the use of an experimental tunnel sprayer for bed grown crops reduced this to 20%. IJzendoorn *et al.* (1998) found on average 51% (27–92%) deposition on the soil surface with conventional 300 l/ha spraying, but spraying 150 l/ha with the use of air assistance deposited 43% (29–63%) on the ground. Thus, the effect of air assistance on soil deposition varies, but in general deposition is lower with air-assistance than with conventional spraying. It is not clear how spray volume affects spray deposition on the ground.

TREE AND BUSH CROPS

Orchards

Table 3.8 summarises the deposition of spray on soil surface underneath apple trees in orchards. When spraying apple trees in full leaf, spray deposition on the ground is on average 25% (18–35%), (Heer and Schut, 1986; Crum and de Heer, 1991; Porskamp *et al.*, 1994*a*, 1994*b*; Ganzelmeier and Osteroth, 1994; Heijne *et al.*, 1995). Spray deposition on the soil is not significantly different between axial and cross-flow sprayers, but tunnel sprayers (Zande *et al.*, 1998) can decrease soil deposition by 50%. When trees are not in leaf, spray

Table 3.7 Spray deposition (% dose) on the soil surface underneath a lily crop (bed-grown) for different periods during the growing season, and different spray techniques[a]

Lily growing season	May	June	(July) August	September	Average
Conventional spray technique	47.4	19.5	19.5	–	29
Air-assistance spray technique	39.8	20.7	18.9	–	26

[a] Spray volume measured in litres/ha.

Table 3.8 Spray deposition (% dose) on the soil surface underneath orchard trees (apple) for different periods during the growing season, and different spray techniques

Apple spray technique	Spray volume (l/ha)	Growth stage		Periods				Average with leaves
		No leaves	Full leaves	Before blossom	After blossom			
					June	July	August	
HV	800		35					35
MV	500		29					29
LV	100–300	60	23	24	25	30	24	25
ULV	<100		25	37	20			27
Tunnel sprayer	100	36	12	14	10	12	6	11
Average conventional		60	28	30	23	30	24	27
Total average		48	25	25	18	21	15	21

deposition on the soil is three times higher than when the trees are sprayed in full leaf (Ganzelmeier and Osteroth, 1994). de Heer and Schut (1986) found 65% more soil deposition before blossom stage than after blossom.

Table 3.9 summarises the spray deposition on apple trees. Average deposition on leaves on apple trees is around 50%. This is the same for conventional (axial and cross-flow) and tunnel sprayers (Baraldi *et al.*, 1993; Porskamp and van der Werken, 1991; Porskamp *et al.*, 1993*b*, 1994*a*, 1994*b*). Stem and branches of apple trees collect on average about 8.3% of the sprayed volume (Herrington *et al.*, 1981). In comparative studies (Metz and Moser, 1987; Raisigl and Felber, 1991) cross-flow sprayers deposit on average 49%, and axial-fan sprayers 39%, on the leaves of the apple trees.

Table 3.9 Spray deposition (% dose) on orchard-tree leaf surface (apple) for different periods during the growing season, and different spray techniques

Apple spray technique	Spray volume (l/ha)	Growth stage		Periods			Average with leaves
		No leaves	Full leaves	After blossom			
				June	July	August	
HV	1000	9	39				39
MV	500	10	48				48
LV	100–300		49	46	70	41	52
ULV	<100	7	52	76	56		62
Tunnel sprayer	100	46	73	28		52	51
Average conventional		8	47	61	63	41	53
Total average		18	53	50	63	46	53

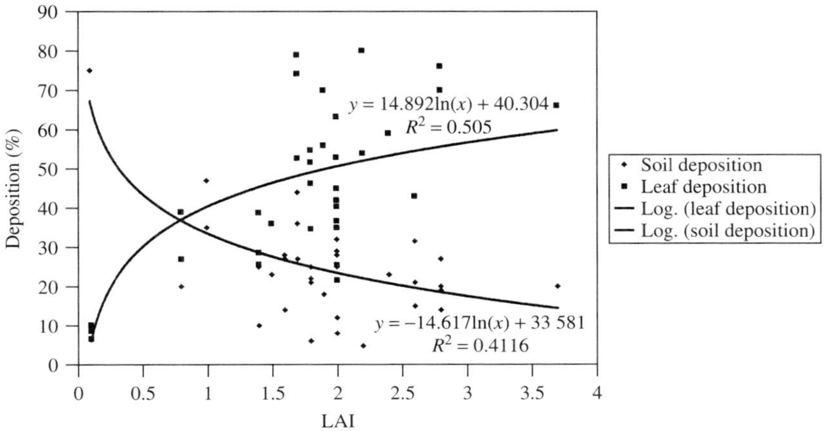

Figure 3.10 Spray deposition on leaf tissue and ground related to LAI in orchard spraying

In orchards, spray volumes range from around 1200 l/ha to 50 l/ha. Spray deposition on apple trees for conventional spraying varies with spray volume. When spraying at 1200 l/ha, the average spray deposition is about 40% (Herrington, 1981), but decreasing the spray volume to 500 l/ha increases deposition to around 48% (Herrington *et al.*, 1981; Metz and Moser, 1987; Cross, 1991; Raisigl and Felber, 1991). Further reductions to low-volume applications (100–300 l/ha) increases average spray deposition to 52% (Baraldi, 1993; Cross, 1991; Heer and Schut, 1986; Porskamp and Michielsen, 1993*b*; Raisigl, 1991). Spraying at ultra-low volumes (less than 100 l/ha) results in a reduction of spray deposition to around 25%.

At the start of the season, when branches are still without leaves, more spray will be deposited on the soil surface than on the crop. Most of the spray will deposit on the crop when the maximum LAI occurs. This diminishes towards the end of the season. From the data found in the literature, these effects are shown in Figure 3.10 for orchard spraying (Zande *et al.*, 1998).

IMPLEMENTING A HIERARCHICAL APPROACH

In order to implement a targeted approach to pesticide application, quality research information is required. Large differences do occur in leaf canopy deposition and deposition on the soil surface underneath crops (Porskamp *et al.*, 1993*a*, 1994*b*), but even in 'spray accountancy studies' (van de Zande *et al.*, 1998; IJzendoorn *et al.*, 1998; Porskamp *et al.*, 1993*a*) total mass is not always balanced. The data presented here must therefore be used with some caution. Also, most data comes from field trials set up to compare spray techniques where well-maintained and often technically advanced equipment is used. Farmer practice might be less advanced. The quality of deposit information could be improved by appropriate modelling. This would also improve the range of situations that could be considered. Although it is well known that meteorological parameters (such as temperature, relative humidity, wind speed, and wind direction) influence spraying, and a large amount of work has been carried out to quantify their influence on spray-drift (Holterman *et al.*, 1997; Miller, 1993), little work has been carried out on their influence on spray deposition. It is encouraging to note that some of the advanced modelling and measurement techniques that have been applied to spray drift are now being applied to the problem of predicting spray deposition in canopies (Walklate *et al.*, 2000). However, even if the quality of the research information is improved, it is still a prime requirement that this information is placed before growers and their advisers in ways that can be readily interpreted. Simple techniques for indicating the quality of spray deposition and the potential for soil contamination and spray drift are required. This would allow operators to make informed choices for equipment use and optimise their application for their particular growing conditions.

REFERENCES

Bals EJ, 1975. The importance of controlled drop application (CDA) in pesticide applications, *Proceedings of the 8th British Insecticide and Fungicide Conference*, pp. 153–160.

Baraldi G, S Bovolenta, F Pezzi and V Rondelli, 1993. Air-assisted tunnel sprayers for orchard and vineyard. *ANPP–BCPC Second International Symposium on Pesticides Application Techniques*, Strasbourg, September 1993, pp. 265–272.

Beukema HP and DE van der Zaag, 1979. *Potato Improvement, Some Factors and Facts*. International Agricultural Centre, Wageningen, The Netherlands, 1979, 224 pp.

Brunskill RT, 1956, Factors effecting the retention of spray droplets on leaves, *Proceedings of the 3rd British Weed Control Conference*, **2**, pp. 593–603.

Cooke BK, PJ Herrington, KG Jones, NM Western, SE Woodley, AC Chapple and EC Hislop, 1985. A comparison of alternative spray in cereals. In: *Symposium on Application and Biology*. BCPC Monograph, No. 28, pp. 299–309.

Cooke BK, EC Hislop, PJ Herrington, NM Western and F Humpherson-Jones, 1990. Air-assisted spraying of arable crops, in relation to deposition, drift and pesticide performance. *Crop Protection* **9(4)**: 303–311.

Cooke BK and EC Hislop, 1993. Spray tracing techniques. In: GA Matthews and EC Hislop (eds). *Application Technology for Crop Protection*. CAB International, Wallingford, England, 1993, 85–100.

Courshee RJ, 1960. Some aspects of the application of insecticides. *Annual Review of Entomology* **5(1960)**: 327–352.

Cross JV, 1991. Deposits on apple leaves from medium volume, low volume and very low volume spray applications with an axial fan sprayer. In: *Air-assisted Spraying in Crop Protection*. BCPC Monograph, No. 46, pp. 263–268.

Crum SJH and H de Heer, 1991. Het effect van twee verschillende spuitmachines op de depositie en luchtconcentratie van deltamethrin in een appelboomgaard. *SC–DLO Rapport 153*, SC–DLO, Wageningen, The Netherlands, 49 pp.

Diepen CA van, J, Wolf H van Keulen and C Rappoldt, 1989. WOFOST: a simulation model of crop production. *Soil Use and Management*, **5(1989)1**, 16–24.

Enfält P, A Enqvist, P Bengtsson and K Alness 1997. The influence of spray distribution and drop size on the dose response of herbicides, *Proceedings of the Brighton Crop Protection Conference – Weeds, 1997*. November 1997, Brighton, UK, 381–390.

Ganzelmeier H and HJ Osteroth, 1994. Sprühgeräte für Raumkulturen – Verlustmindernde Geräte. *Gesunde Pflanzen* **46(7)**: 225–233.

Hack H, H Gall, Th Klemke, R Klose, U Meier, R Stauss and A Witzenberger, 1993. Phänologische Entwicklungsstadien der Kartoffel (*Solanum tuberosum* L.). Codierung und Beschreibung nach der erweiterten BBCH-Skala mit Abbildungen. *Nachrichtenblatt der Deutschen Pflanzenschutzdienst* **45(1993)1**: 11–19.

Heer H de and CJ Schut, 1986. Depositie- en driftmetingen bij conventionele en nieuwe type spuitmachines in de fruitteelt en in de vollegrondsteelten. Voordracht Katholieke Universiteit Leuven 19 maart 1986. TI-K.VIV. Genootsch. Plantenproduktie en Ekosfeer, 34 pp.

Heer H de, CJ Schut, HAJ Porskamp and LM Lumkes, 1985. Depositie- en driftmetingen bij conventionele en nieuwe typen spuitmachines in een tarwe-, spruitkool- en een aardappelgewas. *Gewasbescherming* **16(6)**: 185–197.

Heijne B, H de Putter and J Westerlaken, 1995. Vergelijking Joco en Munckhof tunnelspuit 1994. *Intern Rapport, Proefstation voor de Fruitteelt*, Wilhelminadorp, 9 pp.

Herrington PJ, EC Hislop, NM Western, KG Jones, BK Cooke, SE Woodley and AC Chapple, 1981. Spray factors and fungicidal control of apple powdery mildew. In: ESE Southcombe (ed.), *Application and Biology. Proceedings of a Symposium held at the University of Reading, Berkshire 7th–9th January 1985*. British Crop Protection Council, BCPC Monograph No. 28, Croydon, UK, 1985, 289–298.

Hislop EC, 1983, Crop spraying: the need for scientific data, *SPAN*, **26**, 53–55.

Holterman HJ, JC van de Zande, HAJ Porskamp and JFM Huijsmans, 1997, Modelling spray drift from boom sprayers. *Computers and Electronics in Agriculture*, **19**, 1–22.

IJzendoorn MT van CAM Schouten, AThJ Koster and JC van de Zande, 1998. Reduced use of fungicides in tulip and lily with new spray techniques. *Verslag Nr. 96. Laboratorium voor Bloembollenonderzoek, Proefstation voor de Akkerbouw en de Groenteteelt in de Vollegrond*, Lisse, Lelystad (in Dutch with English summary).

Joyce RJV, Uk S, Parkin CS, 1977, Efficiency in pesticide application, *Pesticide Management and Resistance*, pp. 199–215.

Lefebvre AH, 1993, Droplet production, In: *Application Technology for Crop Protection*, edited by Mathews GA and Hislop EC, CAB International, Wallingford, England, pp. 35–54.

Mathews GA, 1992 Pesticide Application Methods, Longman Scientific & Technical, Harlow, England, p. 405.

Mathews GA and Hislop EC, 1993, *Application Technology for Crop Protection*, CAB International, Wallingford, England, p. 359.

May MJ, 1991. Early studies on spray drift, deposit manipulation and weed control in sugar beet with two air-assisted boom sprayers. In: *Air-assisted Spraying in Crop Protection*. BCPC Monograph, No. 46, 89–96.

Metz N and E Moser, 1987. Bessere Anlagerung-geringere Abdrift. *Möglichkeiten der Pflanzenschutztechnik im Obstbau. Landtechnik* **42(3)**: 104–106.

Miller PCH, 1993. Spray drift and its measurement, In: *Application Technology for Crop Protection*, Mathews GA and Hislop EC (eds), CAB International, Wallingford, England, 101–122.

Miller PCH and Hobson PA, 1991, Methods of creating air-assisted flows for use in conjunction with crop sprayers, In: *Air-Assisted Spraying in Crop Protection*, BCPC Monograph, No. 46, 35–44.

Parkin CS, AJ Gilbert, ESE Southcombe and CJ Marshall 1994. British Crop Protection Council Scheme for the classification of pesticide application equipment by hazard. *Crop Protection* **13(4)**, 281–285.

Porskamp HAJ, and J van der Werken, 1991. Closed Loop spuitsysteem voor het beperken van emissie. *Landbouwmechanisatie* Nr **4**, April 1991, p. 63.

Porskamp HAJ, JMGP Michielsen and C Sonneveld, 1993*a*. Helft reductie met lucht. IMAG–DLO onderzoekt depositie- en emissiemetingen bij de luchtondersteunde veldspuit. *Landbouwmechanisatie* Nr **5**, Mei 1993, 19–21.

Porskamp HAJ and JMPG Michielsen, 1993*b*. Depositie en emissie van de JOCO-tunnelspuit in 1991. IMAG-DLO nota p. **93–72**, IMAG-DLO, Wageningen The Netherlands, 18 pp.

Porskamp HAJ, JMPG Michielsen and JFM Huijsmans, 1994*a*. Emissiebeperkende spuittechnieken voor de fruitteelt (1992). Onderzoek depositie en emissie van gewasbeschermingsmiddelen. *IMAG-DLO Rapport* **94–19**, IMAG-DLO, Wageningen, pp. 43.

Porskamp HAJ, JMPG Michielsen and JFM Huijsmans, 1994*b*. Emissiebeperkende spuittechnieken voor de fruitteelt (1993). Onderzoek depositie en emissie van gewasbeschermingsmiddelen. *IMAG–DLO Rapport* **94–23**, IMAG-DLO, Wageningen, The Netherlands, pp. 33.

Porskamp HAJ, JMGP Michielsen and JC van de Zande, 1997. Driftbeperkende spuittechnieken voor de bloembollen. Drift bij een luchtondersteunde veldspuit, een spuit met een afgeschermde spuitboom en een tunnelspuit voor bedden. *IMAG–DLO Rapport* **97-08**, Instituut voor Milieu- en Agritechniek, Wageningen, The Netherlands, 36 pp.

Raisigl U and H Felber, 1991. Comparison of different mistblowers and volume rates for orchard spraying. In: *Air-Assisted Spraying in Crop Protection*. BCPC Monograph, No. 46, 185–196.

Ringel R, 1991. Wirkung des gelenkten Luftstroms an einer Feldspritze. *Agrartechnik* **41(3)**: 105–107.

Robinson TH, 1993. Large-scale ground-based application techniques, In: *Application Technology for Crop Protection*, Mathews GA and Hislop EC (eds), CAB International, Wallingford, England, pp. 163–186.

Southcombe ESE, PCH Miller, H Ganzelmeier, JC van de Zande, A Miralles and AJ Hewitt, 1997. The international (BCPC) spray classification system including a drift potential factor. *Proceedings of the Brighton Crop Protection Conference – Weeds, 1997*. November 1997. Brighton, UK, 371–380.

Taylor WA and PG Andersen, 1991. Enhancing conventional hydraulic nozzle use with the Twin spray system. In: *Air-Assisted Spraying in Crop Protection*. BCPC Monograph, No. 46, pp. 125–136.

Walklate PJ, Richardson GM, Cross JV and Murray RA, 2000. Relationship between orchard tree crop structure and performance characteristics of an axial fan sprayer, *Aspects of Applied Biology*, **57**, 285–292.

Wretblad, 1997. Foerdelning av sprutvaetska i spannmals- och potatisbestand med fyra olika appliceringstekniker. Spray deposits in cereal and potato canopy with four different application techniques. Report 223, *Department of Agricultural Engineering, Swedish University of Agricultural Sciences*, Uppsala, Sweden, 1997, 20 pp.

Zadoks, JC, TT Chang and CF Konzak, 1974. A decimal code for the growth stages of cereals. *Eucarpia bulletin* **7**, 1974, 42–52.

Zande JC van de, 1995. Deposit measurements and biological efficacy, the effects of volume rates and air assistance on weed control. *Proceedings, Brighton Crop Protection Conference, Weeds, 1995*, pp. 1135–1140.

Zande, JC van de, HAJ Porskamp and JFM Huijsmans, 1994*a*. Air-assisted spraying in potatoes – results of deposition measurements and a biological evaluation. *International Conference on Agricultural Engineering, Milano, 29th August–1st September 1994*, 720–721.

Zande, JC van de, R Meier and MT van IJzendoorn, 1994*b*. Air-assisted spraying in winter wheat – results of deposition measurements and the biological effect of fungicides against leaf and ear diseases. *BCPC Conference – Pests and Diseases Brighton 1994; November 1994*, British Crop Protection Council, Farnham, UK, 313–318.

Zande, JC van de, M van IJzendoorn and R. Meier, 2000. The effect of air assistance, dose and spray interval on late blight control *Phytophthora infestans* in potatoes. *BCPC Conference – Pests and Diseases 2000, Brighton 13–16 November 2000*, British Crop Protection Council, Farnham, UK.

Zande, JC van de, HAJ Porskamp and HJ Holterman, 1998. Spray deposition in crop protection. *Environmental Planning Bureau, Series 5*, DLO Winand Staring Centre, Wageningen, The Netherlands, 1998.

Zande, JC van de, HAJ Porskamp, JMGP Michielsen, MT Van IJzendoorn and R Meier, 1999. Spray deposition and biological efficacy in potatoes. In: J Pabis and RS Rowinski, 1999. *Papers Presented at the Techniques for Plant Protection Conference, TPP, International Conference Warsaw – Poland, 23–26.05.1999*. Warsaw Agriculture University, Warsaw, 117–125.

Zande, JC van de, R Meier and MT van IJzendoorn, 1996. Spraying in-field vegetables: deposit and biological efficacy, the effects of volume rates, dose, spray interval and air assistance on disease control. *BCPC Conference – Pests and Diseases 1996, Brighton 18–21 November 1996*, British Crop Protection Council, Farnham, 343–348.

4 Handling and Dose Control

P.C.H. MILLER
Silsoe Research Institute, Wrest Park, Silsoe, Bedford, MK45 4HS, UK

An important consideration in the formulation and application of pesticides is their handling, packaging and safe use. This includes not only safety to operators but also minimising environmental contamination.

PACKAGING AND CONTAINER DESIGN

FACTORS INFLUENCING THE DESIGN OF PESTICIDE PACKAGING FOR DELIVERY TO THE FARM

The design of containers for delivering plant-protection products to the end user is an important component of the overall process of ensuring that such products are used safely and efficaciously. Gilbert (1998) identified the following factors as being important in designing such containers:

- being able to contain the product during transport, handling and storage with a minimum risk of leakage or spillage;
- resisting physical forces associated with storage, handling, transport, filling and emptying;
- protecting the product from atmospheric degradation;
- providing user safety and convenience during storage, handling, opening, pouring and disposal;
- providing a compatible interface with closed transfer or other loading systems on sprayers.

The type of packaging to be used will be closely related to the physical form of the formulated product. Liquids are normally contained in bottles and cans made of a material that is selected to give chemical resistance, meet storage and handling requirements and provide an appropriate interface for the user. Minimising the risk of operator contamination during pouring has been achieved by using a wider internal neck to reduce glugging and splashing. For such containers, standardised sizes of 1.0, 3.0, 5.0 and 10.0 litre capacities have been agreed, with the smallest

Optimising Pesticide Use Edited by M. Wilson
 ISBN: 0-471-49075-X

1.0 litre size having a 45–50 mm closure, and all other sizes being standardised to the 63 mm closure. Opening containers in a safe way is also an important feature of the design. Results of a study reported by Hancock (1993) indicated that the removal of the foil seal often resulted in high levels of operator contamination. Closures and handles must be such that they can be used effectively with a gloved hand.

Many containers for liquids are made of plastic, and there are then issues relating to the need to provide a barrier between potentially aggressive solvents and the main structural plastic from which the container is made. Hancock (1993) discusses the possible ways in which barrier systems can be incorporated into the container structure, and refers to the use of five- and six-layer structures that will meet a standardised mechanical drop test conducted at a temperature of −18°C. Mechanical strength is also tested when storing solvent-based products for six months at 50°C.

Disposal of used pesticide containers is now a major issue influencing the delivery of such products to the farm. Hibbitt (1998) indicated that some 14 million packs were manufactured in the UK, generating some 3000 tonnes of packaging waste. The need to safely dispose of containers has implications for rinsing, since residues will have consequences for disposal. There has therefore been a marked trend to design containers that can be easily rinsed, by avoiding complex recesses such as in hollow handles or recessed bases. Disposal by burying must be regarded as unacceptable, since many containers are made of materials such as high-density polyethylene that are not biodegradable. Recent research has defined a method for the on-farm burning of containers that minimises smoke, the emission of toxic gases and the generation of ash. A study reported by Goldsworthy and Carter (1998) indicated that the overall environmental consequences of on-farm burning may be more advantageous than first thought, because of the energy requirements associated with any collection and centralised disposal function. However, the need to address disposal issues is one of the drivers that is leading to the development and use of small-volume returnable containers. Such packaging, used in conjunction with closed-transfer systems, provides the opportunity to deliver plant-protection products to the farm with the minimum risk of environmental damage and operator contamination associated with leakage, spillage and container disposal.

The requirements relating to the packaging and transport of plant-protection products will generally relate to:

- the physical form of the product;
- the toxicity profile;
- the sizes of packs.

For handling, packaging and transport, products formulated as water-dispersible granules have important advantages, in that any spillages can be swept up. Packaging using, for example, waxed paper and cardboard, is more amenable to on-farm burning. These advantages must, however, be set against the potential problems associated with the measurement of part packs and small quantities of formulated products on the farm.

The packaging and transport requirements associated with plant-protection products constitute an international issue covered by agreed procedures relating to classification, identification, packaging, marking and labelling, documentation and training. Many of the procedures address issues wider than the delivery of pesticides, and a detailed discussion of these requirements is beyond the scope of this chapter.

RECENT DEVELOPMENTS IN PACKAGING DESIGN

The main factors influencing packaging design can be considered to be the need to reduce disposal requirements, minimising the operator handling of concentrated formulations and the trend towards smaller volumes/weights of formulation required to treat a given area. The development of returnable containers, particularly with some form of metering system to measure material withdrawn and delivery to the application machinery through a closed-transfer system are discussed elsewhere in this chapter.

High levels of operator protection can be achieved by using formulations packaged in a water-soluble bag, such that the complete bag is then put into the sprayer tank or induction system. Not all formulations can be packaged in this way, and the use of a unit package could increase the need for the disposal of dilute pesticide unless some field areas are not treated at full dose. With water-soluble bags, dry storage is obviously essential. The sprayer will need to have a good agitation system to ensure that the packaging is completely dissolved and the formulation dispersed in the tank. Some problems have occurred when loading sprayers with water at low temperatures and with sections of water-soluble bag becoming stuck in the loading system, particularly in induction hoppers.

Formulation as tablets gives similar advantages and disadvantages to those associated with water-soluble bags. Many tablets include the use of an effervescent agent that ensures the rapid dispersion of the formulation. Care is needed when using water-soluble packaging and tablet-type formulations as part of a tank mix, since other components could impede dispersion of the formulation. Under such circumstances, tablets and water-soluble packs should be loaded into the sprayer ahead of other components.

TRANSFERRING THE CHEMICAL FROM ITS PACKAGING INTO THE APPLICATION SYSTEM

The basic means of transferring crop protection products from their original containers is by pouring. For all types of formulation and packaging, there are four main criteria that can be used to characterise the transfer process, namely:

- the time taken to complete the transfer of formulated product required to treat a given area;
- the risk of operator and environmental contamination during transfer;
- the levels of residue that remain in the packaging after primary transfer, and the ease with which these can be removed during a defined rinsing process;

Work rate is a major factor influencing the timeliness of an application. This in turn can influence product performance, both in terms of efficacy and drift, since the ability to select crop, pest/disease pressure and weather conditions at the time of application enables significant changes in product performance to be achieved. The time taken to transfer product is often quoted as a reason for not adopting systems that can reduce the potential for operator and environmental contamination during a transfer operation – see p. 49 and 53. The current trend towards formulating products such that smaller quantities are required to treat a defined area has implications for facilitating rapid transfer, but may also have repercussions relating to the effect of residues and contamination when using more concentrated formulations.

IMPLICATIONS FOR CONTAINER AND PACK DESIGN

A container should enable the contents to be completely emptied in a relatively short time, and the internal surfaces to be rinsed effectively. Smith (1998) indicated that the best containers would therefore have the features of a wide 'funnel'-shaped neck, and no residue trapping areas such as are associated with some designs of hollow handle. The use of a wide neck and standardised 63 mm opening can facilitate the pouring of liquid formulations with the minimum of the glugging and splashing that have been shown to be responsible for high-levels of operator contamination (Gilbert, 1998). The use of a standardised closure also has important implications for the design and development of closed-transfer systems – see p. 53 of this Chapter.

Dispensing part of the contents of a pack gives a need to be able to assess the contents of a partially used pack for both dispensing, audit, and stock control reasons, and to be able to reseal the pack once a quantity of product has been removed. For liquid formulations, resealing is commonly via a screw cap with a waxed paper/plastic washer fitted within the cap. Graduations on the side of clear or translucent containers enable estimates of the contents to be made and, in some circumstances, may be adequate for measuring quantities of product required to treat a defined area with a given dose. More generally, some method of measuring part-packs when pouring will be required. For liquid formulations, this will commonly involve the use of a calibrated volumetric measuring cylinder or jug. These are relatively low cost and robust, and therefore well suited for use in an on-farm chemical store. Formulations such as water-dispersible granules (WDGs) are more difficult to measure volumetrically, because of variations in packing density, but few farm stores are equipped to weigh out part-pack quantities of such formulations. Packaging quantities to treat defined, small areas at full-dose (e.g. hectare packs) can aid the handling of such formulation types, but can then lead to a time-consuming process when loading large sprayers operating at low volume application rates. The use of part-packs of highly active formulations can also involve the measurement of small quantities of all types of formulation, and this can be difficult to achieve with acceptable accuracy in many on-farm stores. The use of secondary chambers within a pack for measuring purposes is used in the domestic garden markets, but is not popular in the agrochemical sector, mainly because of the problems associated with the rinsing and cleaning of such containers.

For containers emptied by manual pouring, a triple rinsing technique has been adopted as the standard for comparing different types of container, formulation and alternative rinsing methods. In Europe, the Dutch established a covenant that required containers to be rinsed, such that less than 0.01% of the original contents remained after rinsing. Smith (1998) indicated that this level of cleanliness was determined pragmatically as a threshold at which used pesticide containers could be allowed to enter the municipal waste stream with no direct link to the toxicity of the formulation. A number of studies have shown that triple rinsing procedures can give residue levels that meet the 0.01% requirement. Lavers (1993) measured residues in a 5.0 litre container filled with a commercial formulation ('Bladex') that were rinsed in different ways, and showed that manual triple rinsing gave the lowest residue levels at 0.0008% of the original contents of the container. Smith (1998) quoted results from a survey of agricultural chemical company tests, using different formulations and container designs, that indicated that the triple rinsing procedure could give residues less than the 0.01% requirement in more than 92% of cases. However, it has also been recognised that the triple rinsing procedure takes time to implement: Gussin (1998) reporting an estimated time of approximately 112 seconds to triple-rinse a 5-litre container. Both the time taken and level of cleanliness will be a function of container design and the properties of the formulation.

A number of semi-automated systems for cleaning containers have been developed, and these are commonly operated in conjunction with induction hoppers – see below.

INDUCTION HOPPERS AND OTHER PESTICIDE LOADING SYSTEMS

A series of studies, including work conducted by the UK Health and Safety Executive in the 1980s (Anon., 1986; Frost and Miller, 1988), identified a high-level of risk of operator and environmental contamination associated with the filling of agricultural sprayers. This was particularly the case when access to the filling hole in the main tank required the operator to climb on to the top of the tank carrying an opened pesticide container. As the size of spraying machinery has increased, so the practicality of an operator making repeat visits to the top of the tank during the loading operation has become less acceptable, because of the manual effort and time involved, as well as the contamination risks. A number of systems have therefore been developed that enable the operator to load the sprayer with pesticide formulations while standing on the ground.

The low-level filling system that has become the most popular in Northern Europe is the induction hopper. This commonly comprises a bowl of approximately 0.5 m dimension into which chemical is poured manually. Most bowls now use a Venturi valve arrangement positioned at the base of the hopper to suck material from the bowl into a liquid flow, that is then directed to the main sprayer tank. An alternative direct connection of the induction hopper to the input side of the main pump is much less popular, because of the potential to damage the pump from extraneous materials that could be introduced via the hopper. A typical induction hopper design will therefore comprise:

- the bowl structure with a connection to the main sprayer circuit via a Venturi valve;
- a valve for controlling the flow of material out of the hopper;
- a system for delivering a flow of wash-down water to the sides of the hopper;
- a container rinsing system.

A typical layout of an induction hopper system is shown in Figure 4.1.

The use of an induction hopper was recognised as providing a substantial improvement over pouring chemical directly into the main tank with regard to potential operator and environmental contamination, but that a poorly designed and installed hopper would either not be used or could create a high-level of contamination risk. A British Standard was therefore prepared (Anon., 1996*a*), detailing both a physical and performance specification for induction hoppers operated in conjunction with crop sprayers. The standard accounts for induction hoppers being designed to operate with liquid, granular and powder formulations and specifies performance in terms of chemical resistance, loading rates that can be achieved, potential for operator and environmental contamination and residues in containers rinsed using the system. To comply with the standard, the hopper should have the following main physical features:

- a minimum volume of 15 litres,
- a minimum filling hole dimension of 250 mm;
- a loading height of between 500 and 1000 mm above ground level, with a clearance of 500 mm above the hopper;
- instructions for use supplied;
- clearly labelled controls;
- a clearance dimension in the bottom of the hopper of less than 10 mm (with a mesh strainer fitted if necessary) to prevent debris from being drawn into the sprayer circuit.

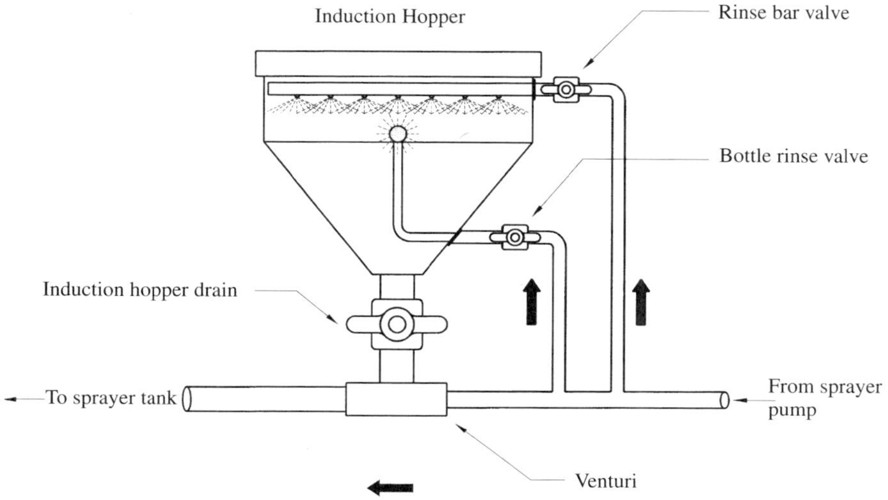

(a)

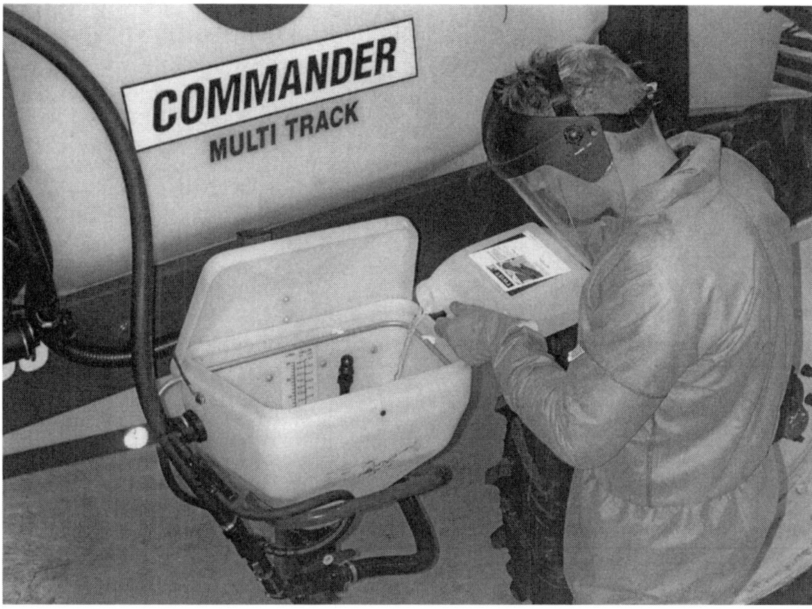

(b)

Figure 4.1 A typical layout of an induction hopper system. (a) Flow circuit diagram. (b) typical installation on a sprayer

The performance requirements for a hopper to comply with the standard relate to:

- *chemical resistance* – all components that make up the hopper must be able to resist any chemical action, and this is tested using an aggressive test liquid with components typically immersed for a 12-hour period – to comply, changes in weight or any dimension over the test period must be less than 5% of the original;
- *chemical loading rate* – a minimum rate is specified because it was recognised that systems that were very slow would not be used in practice – the standard requires a minimum rate of 12 litres/min for a test liquid formulation and 6 kg/min for test granule and powder formulations measured over an operating period of at least three minutes;
- *leakage and potential operator contamination* – with the induction hopper operating at the maximum transfer rate and with a full flow of wash-down water, the maximum leakage and losses due to splashing shall not exceed 1.0 ml when sampled over a three-minute operating period;
- *residues within the hopper* – on completion of a transfer operation, the total residue on internal and external surfaces must be less than 1.0 ml, following a transfer operation and rinsing procedure as defined by the supplier of the hopper;
- *residues within a container* – the induction hopper must be fitted with a device that will rinse containers such that the maximum residue following a defined rinsing procedure shall be less than 0.01% of the volume of the container – this to be verified in tests with the test liquid formulation.

Standard test methods have been developed as part of the British Standard, using defined test liquid, granular and powdered formulations. A number of commercially available units are now able to claim compliance with the British Standard. A key component in making such a claim is a definition of the operating parameters for the hopper, particularly relating to the range of container sizes for which the unit is designed to operate and the instructions for use, including the container rinsing system.

Container rinsing systems designed for use in conjunction with induction hoppers can be considered as being one of two basic types, namely:

- a pressure jet type in which a number of jets of water are directed into the container; and
- a rotating nozzle type where cleaning liquid is directed at the internal surfaces of a container by a rotating element powered by the liquid flow.

Rotating systems often have an advantage of improved coverage of cleaning water contact with the inside surfaces of the container, but may also give less momentum at the container surface. A high-momentum flow, as associated with some jet arrangements, gives good cleaning performance, but makes movement of the container over the rinsing nozzle an important parameter when defining cleaning performance. While it has been shown that most operators are capable of achieving acceptable residue levels when using a pressure-jet system (Cooper and Taylor, 1998), few published results are available comparing the performance of different designs of container rinsing system. Studies have generally shown that a larger number of short rinsing cycles give a better cleaning performance than a smaller number of larger cycles. This is particularly true when the flow of rinse liquid out of the container is impeded by the mounting and supply systems associated with the rinse nozzle, such that there is an accumulation of water in the neck region of the container.

Many tests of container rinsing performance associated with induction hoppers are conducted with the systems operating with clean water, whereas for many installations on practical sprayers the rinsing of the container uses liquid from the main tank. It is likely that the differences in residue levels associated with rinsing with clean water or tank contents will depend on the nature of the tank mix formulations, as well as on the design of the hopper and the container rinsing system.

An alternative to the induction hopper system that facilitates the low-level filling of larger sprayers is the induction probe (Frost and Miller, 1988). As with induction hoppers, probes can be connected to either the inlet side of the main sprayer pump, or, using a Venturi valve, to the outlet pressure side. Probes have been designed that will handle both liquids and powders in the dry state, provided that the main probe tube is kept dry prior to any transfer operation. Although probe designs enable filling from ground level and eliminate the need to pour liquid formulations, the systems commonly installed on machines in Europe did not constitute a closed-transfer system.

When withdrawn from a container during a transfer operation, there is a high risk of operator and environmental contamination from:

- quantities of formulation remaining within the probe tube that run or drip out of the exposed tube; and/or
- there is a high risk of contamination from residues on the outer surface of the probe tube.

Probes have been designed that can overcome most of such risks and therefore can be regarded as closed-transfer systems, but these are relatively costly, complex and slower in operation than a simple 'dip' probe.

CLOSED-TRANSFER SYSTEMS

Recognising that the potential for direct operator contact with the concentrated formulation, particularly liquids, during the loading of application equipment constituted one of the highest risk factors relating to operator and environmental contamination led to the development of closed-transfer systems. The main impetus for system development came from the specification to use closed-transfer systems in regulations in the USA. In 1974 the California Department of Food and Agriculture introduced regulations that required a closed system to be used for transferring liquid pesticides in the most toxic category from their container into a sprayer. At the time when the regulations were introduced, no closed-transfer systems were commercially available, and enforcement was postponed until 1978 to allow systems to be developed. A number of systems were then developed (Frost and Miller, 1988) to enable users to meet the regulations. In the UK it was not until the 1990s that commercial efforts were directed towards producing closed-transfer systems for use with agricultural crop sprayers.

The main types of system are listed in Table 4.1 together with the main features of each system. For a probe arrangement to become a closed-transfer system it is necessary to protect the portion of the dip tube that is exposed to the chemical formulation. There is also a need to make a seal between the probe and the container, with a valve arrangement to enable air to enter the container as the formulation is drawn out. This can be done by using a probe with a built-in container closure and air-bleed valve arrangement, together with a concertina-type shroud that covers the main part of the probe dip tube as it is removed from a container. This type of probe system can be cleaned by drawing clean water through the system, either from the original or a second container. The system does not normally enable the original container to be rinsed/washed using the probe. The probe system has not proved to be commercially popular in Europe and the UK, being relatively cumbersome and slow to use.

Container puncturing systems have the advantage, in principle, of being able to operate with a wide range of packaging and formulation types, although products in glass packaging cannot be handled satisfactorily. Particular systems may only be able to operate with a limited range of container sizes and shapes. Systems generally comprise a vessel that is connected to the sprayer, such that it can be supplied with rinsing water under pressure, and the liquid in the vessel can be drawn into the main tank of the sprayer (Figure 4.2). Full containers are loaded into the vessel, constrained and punctured by a manually operated spike. The spikes are arranged such that rinse water can be pumped through them, into the punctured container, such that main contents and residues are flushed from the containers and into the sprayer. This type of system cannot be used to transfer part contents of a pack, although in some designs it is possible to retain some chemical in the main vessel for use in separate sprayer loads. Systems are commonly bulky, and the unit developed in the UK (the Schering 'Packman') stood separately at the filling site and was not carried on the sprayer. Empty containers recovered after a delivery cycle cannot be reused for any purpose, and should be rinsed to a level such that they can be safely disposed of (residues <0.01% of original container contents). Some units developed in the USA have incorporated a 'can crusher' to reduce the volume of waste or subsequent disposal. Evaluations of the performance of this type of system have shown that they are able to work effectively when operated with packaging types and sizes for which they were

Table 4.1 Types of closed transfer system developed for handling liquid pesticide formulations

Type	Containers that can be used with the system	Examples	Features/comments
Suction probe.	Potentially any but may be cumbersome with small single trip containers. Requires special closure.	None commercially in UK and Europe. 'ChemProbe' in USA.	Position and handling of probe shroud is critical. May not fail safe. Not good for rinsing the pesticide container.
Container puncturing.	Good for plastics, metal and cardboard containers. Not suitable for glass.	'Schering Packman' – not now commercially available. Some systems developed for US market.	Relatively large system often not carried on the sprayer. Punctured containers rinsed ready for disposal. Can be used with all types of formulation but not part packs.
Systems based on standard single trip containers.	Potentially any. Systems often rely on presence of a foil seal so as to remain "closed" while an alternative closure is fitted.	'Wisdom' system.	Contents withdrawn in measured amounts. Container inverted in system. Can also be used with keg type containers.
Specialist low volume systems.	Formulation supplied to end user in a ready to apply form – no dilution necessary.	'Nomix' hand lance system.	Packaging of 'bag in a box' type. Main applications in the local authority/amenity sector.
Systems based on specialist small returnable containers.	Specialised container with closure that forms one half of a dry break coupling. Containers are returned for re-filling.	'CIBA LINK' or LINK-PAC Sotera system – still at prototype stage.	Emptied/transfer by inverting container. Possible to transfer part contents of a container but without direct measurement. Returnable containers are not rinsed.
Systems based on specialist keg type returnable containers.	Specialised returnable kegs. Fitted with dip tube and components for interface valve with tamper-proof seals. Kegs can be tracked and labelled electronically.	'Ecomatic – Cyanamid Signet 2000' – Team Sprayers. 'Wisdom' system with 'Micromatic drum valve'.	The development of a 'standard' interface valve has been a key part of such developments. Returnable containers are not rinsed but transfer system is.

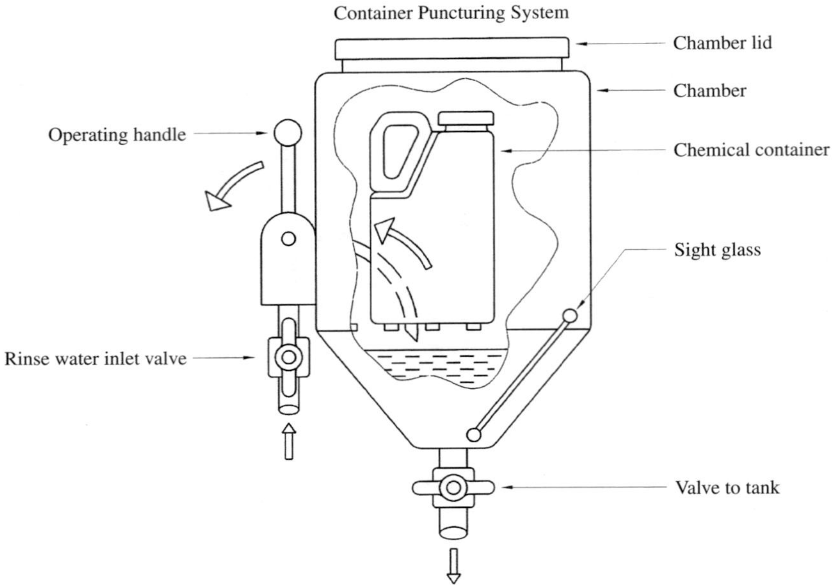

Figure 4.2 A typical layout of a closed-transfer system using the container puncturing principle

designed and that, for some designs, this can be limiting. Some packaging systems using, for example, plastic liners in metal containers, can give problems if both outer and inner materials are not completely punctured (Brazelton *et al.*, 1980).

Hand-held rotary atomiser application systems are able, in some circumstances, to operate at volume application rates that are low enough to enable operation with formulations that do not require dilution prior to use. In this case, the container of formulated product can be connected directly to the application system using an arrangement that minimises any loss or dripping when full containers are loaded or empty ones removed.

In the UK, Wisdom Agricultural Ltd developed a system that can be used with small single-trip containers as well as with larger kegs (Felber, 1993). The main component in the system is effectively a large syringe that draws a measured quantity of liquid formulations from the container and delivers it to the sprayer circuit. When operated with small (5–10 litre) single-trip containers, the container cap is removed and replaced with a specialist unit incorporating two dry-break couplings and an arrangement that breaks the foil seal as a new seal is made with the container. The dry-break couplings enable the container to be connected to the syringe-based system, the container to be inverted and a measured volume of formulation to be delivered to the sprayer. The use of two connections enables air to be introduced as chemical is withdrawn through a riser tube, such that the pressure inside the container is always close to atmospheric. The adoption of a standardised 63-mm opening and specified thread for containers was an important step that enabled closed-transfer system designs to interface with containers. When used with small-volume refillable containers (kegs), then the connection to the system is via a valve arrangement, similar to that described by Perryman (1993). This type of valve consists of two main components: an extractor valve that is integrated within the container and includes a dip tube and valve arrangement, such that when connected to the coupler, the other main component, liquid, can be drawn from the container while air is also drawn in. The valve/coupler interface is then a key part of this and other closed-transfer systems that are based on the use of small-volume refillable containers, such that there are proposals to adopt the form of design shown in Figure 4.3 as a

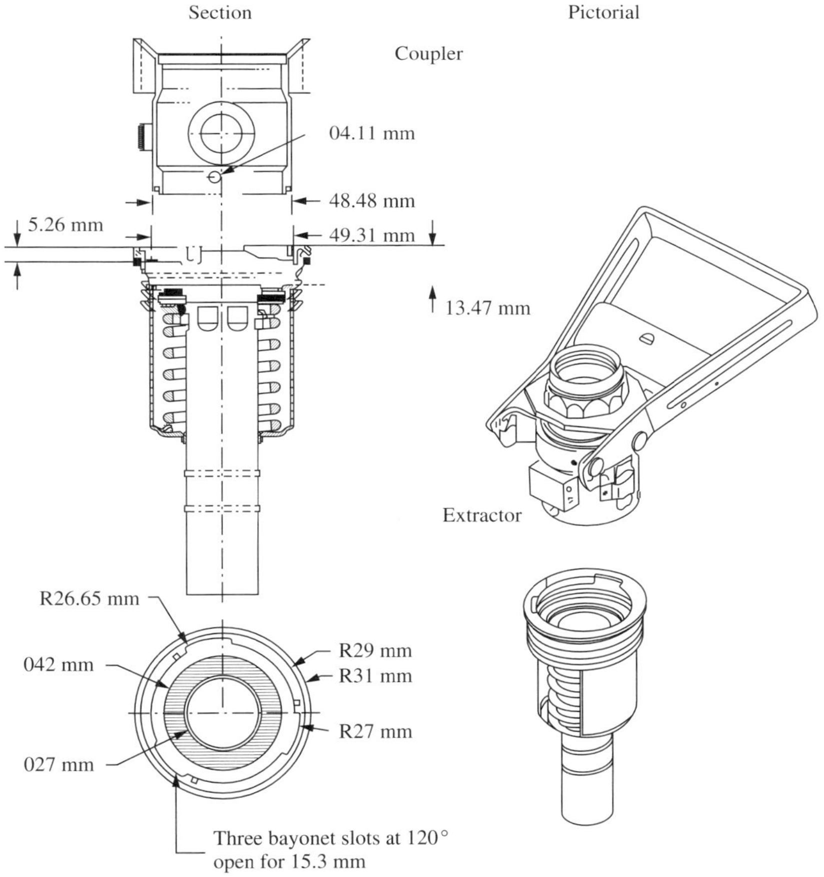

Figure 4.3 Valve/coupler arrangement proposed as a standard interface for closed-transfer systems using small-volume refillable containers

standard interface. Constructed in stainless steel and with Viton seals to resist chemical action, this valve/coupler system is now used in a number of designs of closed-transfer system employing different methods of transferring the liquid and measuring the amount of liquid that has been transferred.

A purpose-designed 10-litre graduated refillable plastic container with two handles and a valve instead of a sealed closure is the basis of a closed-transfer system, the CIBA–LINK, developed in conjunction with a major agricultural chemical manufacturer (Felber, 1993) and now known as the LINK–PAC, as shown in Figure 4.4. The system is used by initially installing a female coupling unit in an appropriate part of a sprayer such as the lid of a conventional induction hopper. To transfer product, the container is inverted and the valve on the container mated with a coupler on the sprayer. The valve/coupler arrangement is opened by rotating the container and liquid then flows under gravity. Part contents of the container can be transferred by monitoring the flow and disconnecting once the required volume has been delivered. When empty, the containers are returned to the original chemical supplier to be checked and refilled.

In the same way that it was recognised that a poorly designed induction hopper would not reduce the risks of operator and environmental contamination (see p. 49) and that a standard relating to system specification and performance was needed (Garnett, 1993) a standard relating to closed-transfer systems for liquid plant protection products has also been produced (Anon., 1996*b*). This follows the same format as the standard relating to induction hoppers and indicates that in order to comply with the standard, a closed-transfer system must

Figure 4.4 The LINK–PAC closed-transfer system

have a specification that includes:

- the ability to interface with containers via a secure and leak-proof connection, with the term 'leak-proof' defined in relation to defined test criteria;
- the ability to rinse both the transfer system and, except when handling returnable containers, the empty container, such that residues in the container are less than 0.01% of original contents;
- operation of the system with clearly labelled components that are between 500 and 1000 mm above ground level and that can be operated with a gloved hand involving a force of less than 50 N;
- not subjecting the original pesticide container to pressures of more than 25 kN/m^2 above atmospheric or 50 kN/m^2 below, and not pressuring the liquid contents by more than 100 kN/m^2 above atmospheric pressure unless secondary protection measures have been put in place or the system has been pressure-tested to five times this value;
- the ability for all components to resist chemical attack as verified in a defined test procedure;
- the provision of clear instructions and a specification for the range of sizes of container for which the system will operate.

Within the standard, the ability to measure the quantities of liquid formulation delivered; to transfer part-contents of a container; and to return undelivered liquid back to the original container, are specified as optional features. This means that systems based on puncturing the containers or transferring the contents in a single batch can meet the standard. The development of the standardised 63-mm neck and thread on containers has been an important step. Many closed-transfer systems that operate with small single-trip containers rely on the pressure of a foiled seal to enable an outer cap to be removed and replaced with a connection to the system. The presence of the remnants of a foil cap can, however, lead to relatively high residues when containers are rinsed by washing systems, because of small quantities of the original formulation that become trapped behind the remnants of the seal. This phenomenon is relevant to rinsing in closed-transfer systems, and also to container-rinsing systems fitted to induction bowls. For this reason, some manufacturers are not now fitting foil seals, and this has important implications for the design and operation of closed-transfer systems, particularly for use in conjunction with smaller pack sizes.

The performance requirements specified within the British Standard for closed-transfer systems for liquid plant-protection products relate to:

- a maximum leakage during any transfer operation of less than 0.25 ml of formulated product;
- a maximum residue on all parts of the coupling following a disconnection of less than 1.0 ml;
- a residue in any part of the system after a defined flushing procedure of less than 0.5 m of transferred formulation.

There are now a number of complete closed-transfer systems commercially available in the UK that are able to meet the requirements of the British Standard, and yet the number of units used on farms is still relatively low. Costs, complexity and speed of operation have been cited as reasons for the slow commercial uptake of such systems. Some chemical suppliers have promoted the use of closed transfer, and the requirement to use a closed-transfer system has now been included as part of the conditions of use for some plant-protection product formulations justified by the toxicological profile of the active ingredients. It is therefore likely that the use of such systems will increase in the future.

METHODS OF CONTROLLING DELIVERED DOSE

Agricultural crop sprayers employ one of two methods for controlling the dose of a spraying system travelling at a defined or measured forward speed, namely:

1. adjustment of the volume output rate, or
2. adjustment of the concentration of plant-protection products within the delivered spray liquid.

Many of the larger sizes of boom sprayer used to treat arable crops in Northern European conditions have a control system that adjusts the volume output rate depending upon measured forward speed, so as to keep the delivered dose constant over a range of operating speeds.

CONVENTIONAL DOSE CONTROL BASED ON VOLUME RATE

Pressure-Control Systems

For a fixed-orifice conventional hydraulic pressure nozzle operating with a given spray liquid, the only way in which flow rate through the nozzle can be changed is by varying the pressure at the nozzle. However, variations in operating pressure result in changes to both the spray volume distribution pattern (patternation), the droplet size (spray quality) and velocity distributions (see Table 4.2).

Table 4.2 Changes in physical spray characteristics from a conventional nozzle operating at different pressures

Pressure (bar)	Droplet size distribution	
	Volume median diameter μm	% volume in droplets <100 μm
1.0	234	5.6
2.0	186	12.0
3.0	167	16.1
4.0	157	19.6

Since flow rate is proportional to the square root of pressure (e.g. to double the flow rate requires a fourfold increase in pressure), the range over which such control strategies can be operated effectively is commonly limited to a nominal value of ±25% (Paice *et al.*, 1996). While this range is often adequate to compensate for changes in forward speed due to differences, for example, in field terrain, it is not adequate for making spatially variable applications of plant-protection products.

Pressure-control systems are based on measurements of forward speed, and either the measurements and control of pressure at the delivery nozzles, or the flow to the boom or boom sections. These two approaches are referred to as either pressure or flow control, depending on the sensing system used, but the method of control is the same in each case i.e. adjustment of pressure at the delivery nozzle. Control based on a measurement of flow has the advantage of giving a direct indication of delivered spray volume rather than an assumed relationship between pressure and flow through a nozzle. However, flow measurements can pose difficulties, particularly regarding operation with a range of liquids and in terms of calibration and reliability. The performance of pressure gauges can also be checked by conventional calibration procedures, and care is needed to ensure that the measured pressures are the same as at the spray nozzle.

Pressure is commonly controlled in systems using diaphragm or piston pumps by adjusting the proportion of the output flow from the pump that is diverted into a return line back to tank. This can be actuated by a simple butterfly valve, an adjustable spring-loaded ball valve, or a valve based on the use of air pressure to control the line pressure. Measurements of the response characteristics of commercially available valve systems have shown that most are able to achieve within 10% of a set value in less than 10 s (Rietz *et al.*,1997), and this then has formed the basis of environmental performance standards for agricultural crop sprayers that have recently been agreed within the European Community.

Although pressure control is the most common way of adjusting the volume output of boom sprayers fitted with conventional nozzles, two other main approaches have also been used, namely: the use of twin-fluid nozzles and arrangements based on the use of solenoid valves. It should also be noted that the use of other types of spray generation system, such as spinning discs or air shear nozzles give other options for control of applied volume rate, but such arrangements are not commonly used on boom sprayers.

Twin-fluid Nozzle Systems

In this type of nozzle, spray liquid and air are mixed together as they pass through chambers in the nozzle body to form a spray that is delivered from a modified flood or reflex type nozzle. The flow of air contributes directly to the spray droplet formation process. By controlling air and liquid pressures, liquid flow rate and spray quality from the nozzle can be varied independently. Droplets generated from such nozzle designs can contain air inclusions if they are above

100 μm in diameter (Miller *et al.*, 1991), and hence determination of spray quality cannot be directly comparable with the techniques used for conventional hydraulic pressure nozzles.

The adjustment of nozzle flow rate at given spray quality by altering both air and liquid pressures has been used commercially in automatic control systems to compensate for forward speed changes with this type of spraying system. Commercially available performance data indicate that this type of system can give a range of volume flow rates of up to 2.5 : 1 when operating to produce a defined spray quality. The response times of this type of system are directly comparable with those of conventional pressure-control system, since the actuating valves to control liquid and air pressures are the same in both cases. It should be noted, however, that the output flow characteristics of a twin-fluid nozzle are more sensitive to changes in pressure than is a conventional system, and this may lead to a more rapid response characteristic.

Systems Based on Solenoid Valves

Solenoid valves are a common way of controlling the flow of spray liquid as an on/off control on crop sprayers. A range of volume rates can therefore be achieved by using multiple nozzle lines on a boom and using solenoid valves to control the supply to each set of nozzles. Such an approach has the disadvantages of relatively high cost and only being able to operate at discrete levels of delivery (Paice *et al.*, 1996), with the steps between levels reduced by the use of more supply lines.

Recent commercial developments have sought to exploit an arrangement originally proposed by Giles and Comino (1990), in which the output from a nozzle is pulsed rapidly. In the original work, a fast-acting solenoid valve was placed immediately upstream of each nozzle. The solenoid coil was excited with a rectangular wave signal with a repetition rate of approximately 10 Hz and using a variable duty cycle of 10% to 70%. The valve switched on and off completely at this frequency, and over this duty cycle range. Therefore, for most of the period of the driving waveform, the pressure at the nozzle inlet was either at zero or at a value well within the nozzle's optimum range.

This technique was shown to have significantly less effect on spray droplet size and spray pattern than is typical of pressure-control systems. Response time was as fast as that achieved with standard on/off solenoid valves as described above, but the maximum flow-rate range that could be achieved was limited by the switching time of the valve. Experimental results showed that a range of 3 : 1 could typically be achieved.

INJECTION METERING SYSTEMS

The concept of injection metering, in which the concentrated chemical formulation is mixed with the diluent, usually water, is attractive because it has the potential to substantially reduce the need for decontaminating the sprayer and disposing of unused dilute formulation, including washings, and minimising operator contamination particularly in relation to maintenance tasks. If developed in conjunction with the use of closed-transfer systems, then operator contamination associated with the loading and mixing operations would also be substantially reduced (Frost and Miller, 1988). Landers (1993) evaluated the performance of a commercial design of injection metering system, and considered the financial and environmental advantages of reducing tank residues. He quoted tank residue levels, (Table 4.3), which were between 2.2 and 4.2% of tank contents. The performance characteristics of a spraying system using injection metering are dependent upon the detailed configuration used and particularly the position of the injection point and the control strategy employed.

Injection can take place either up or downstream of the main system pump. For injection on the upstream, low-pressure input side of the main pump, the injection system has to operate against a relatively low delivery head, and this means that metering pump designs based on a peristaltic pump can be used. However, since all of the output from the main pump does not go directly to the nozzles, because of the need to control delivery pressure, there is a need to incorporate a recirculation loop, such that dilute formulation not delivered to the nozzles is returned to the input side of the pump rather than the tank, as in a conventional pressure-control configuration. Injection on the downstream, high-pressure side of the main pump normally takes place after the pressure-control system, so that clean water is returned to the tank by the pressure-control system. The metering pump must then deliver against the spraying pressure, and this means that piston type pumps are needed. Injection on the pressure side of the circuit could take place at any point between the pressure-control system and the nozzles. However, since a metered flow of concentrated formulation cannot be divided without losing control, injection downstream of the controls for each boom section would involve using more metering pumps and an associated increase in costs. In the limit, it has been proposed that the formulation should be injected at each nozzle, but this gives problems relating to the number and costs of metering pumps and the lengths of pipe that would be carrying concentrated formulation. Design experience has shown that for a high-pressure injection system it is desirable to keep the lengths of delivery pipe as short as possible, so as to minimise safety concerns relating to the transport of formulated product under pressure and difficulties associated with flushing and cleaning on completion of a spraying operation (Paice *et al.*, 1993).

Table 4.3 Residual volumes in conventional crop sprayers (after Landers, 1993)

Machine	Machine type	Tank size (litres)	Boom size (m)	Residual volume (litres)
A	Trailed	2000	24	83.3
B	Trailed	1500	18	40.5
C	Mounted	625	2	13.5

An injection metering system can be operated based on one of two control strategies, namely:

- the use of a constant flow of water (diluent) with injection to vary the concentration of formulated product in the spray liquid and hence control delivered dose – this has the advantage that the nozzle system and pressure can be selected to give the optimum spray quality and distribution pattern, that are not then changed by the need to adjust sprayer output; or
- the use of a constant concentration strategy, whereby the output of the sprayer is controlled by adjusting both the volume rate and the metered flow rate.

The performance of an injection metering system has components relating to the accuracy of steady-state delivery, the response time to a change in demand, the range of dose rate delivery over which it can operate, and residue/system flushing problems. Studies by Antuniassi *et al.* (1997) showed that the ability to achieve a given performance in terms of steady-state delivery was a function of pump selection and detail design parameters, rather than the control strategies employed. Some direct calibration of metering pump performance is often required to accommodate the wide range of flow characteristics found with liquid plant-protection product formulations. Response times, however, are dependant upon the type of configuration used. A system using a constant flow-rate control strategy and injecting upstream of the main pump will inherently have a long and variable response characteristic compared with injection downstream of the main pump, because of delays in the plumbing system. The use of a constant concentration strategy will reduce delays arising from the pumping arrangements, but the range of delivery rates over which control can be achieved is limited to that of a conventional volume-rate control system. Response times therefore have components relating to both the plumbing arrangements and also the dynamic response system of the metering system. Frost (1990) discusses how plumbing components can be varied in practical system designs, and work by Paice *et al.* (1995) showed that boom plumbing delays could be reduced to less than 5 s by appropriate pipe selection for a variable concentration system injecting on the pressure side of the main pump. The system developed by Paice *et al.* (1995) was shown to have a closed-loop response characteristic that was of an exponential form with a time constant of 1.3 s, and measurements reported by Antuniassi *et al.* (1997) indicate that this is relatively fast in comparison to some commercially available systems.

All metering pumps for use with injection systems must be able to resist chemical attack from solvents used in formulating plant-protection products. They must also be able to operate with a wide range of physical properties, since liquid formulations have a wide range of viscosity and flow characteristics. In practice, this commonly involves the use of high-specification seals within pumps, replaceable seals or tubes in a peristaltic design, with associated implications for complexity and costs. Frost (1990) describes a design, in which formulated product is delivered from a metering cylinder by the action of a gear pump operating to deliver water to the base of a cylinder containing a piston which then displaces the active formulation. Since the metering pump is always working in water, it can be characterised such that it acts as both pump and flow meter in a closed-loop control system. Problems of chemical attack are minimised, and, since the whole of the metering cylinder and the pump delivery loop are at line pressure, the metering pump is only working to overcome the effects of friction in the circuit. The system can therefore operate with formulations having a wide range of flow characteristics without a need for direction calibration. To date, this system has only been developed to the commercial prototype stage.

One of the most important factors that has limited the commercial development of injection metering systems is the wide range of volume rates of formulated product that equate to a prescribed dose. Pump systems can typically achieve a good metering performance over a 10 : 1 range of delivered volume rates, and while this can be extended to more than 25 : 1 by using high-specification components and control systems, this still does not meet the requirements of an injection metering system on an agricultural crop sprayer. The trend towards the development of highly active products now means that a full-dose treatment may involve the application of less than 250 ml/ha of formulated product, while the requirement to meter up to 5.0 l/ha also still exists – a range of 20 : 1. When combined with the need to be able to control applied doses over a range of forward speeds, and with different sections of the boom operational, a range of product flows in excess of 300 : 1 is quickly reached, and this cannot be achieved with a single size of pump. The need to be able to meter the components of typical tank mixes and handle the range of volume flow rates means that most practical systems will involve at least three pumps, and this has implications for system costs. Most chemical metering systems have been designed to operate with liquid formulations. However, recent formulation trends have increased the number of products that are available as water-dispersable granules, because of the safety implications when handling any spillage of such materials. Although prototype systems have been developed for metering granular formulations into both the up- and downstream sides of the main pump, practical problems of metering granules into a pressurised flow of liquid have meant that, to date, the only commercial systems for metering granules operate on the input low-pressure side of the pump. Granules are metered using a development of the fluted roller system, dissolved in a secondary flow of liquid and then pumped into the main feed to the pump with a peristaltic unit.

METHODS OF MATCHING DOSE TO TARGET REQUIREMENTS

The initial development of either volume-based or injection metering dose-control systems aimed at achieving a uniform controlled dose across a treatment area that was independent of machine variables, particularly forward speed. However, there is substantial evidence to show that the target requirements for many plant-protection products are not uniform across a field, and that there is the potential to reduce chemical use rates if applications can be better matched to target requirements. In the UK, the strongest pressure to improve the match between target requirements and application inputs relates to the treatment of grass weeds in cereal crops, for the reasons outlined below:

1. There is some evidence to show that the distribution of weeds such as black-grass and wild oats is commonly not uniform across fields. A study of the distribution of black-grass over a ten-year period reported by Wilson and Brain (1991) concluded that the weed distribution was patchy, and that the patches were relatively stable. It should, however, be noted that this study used a coarse sampling grid, and that small changes in weed patch behaviour may not have been detected. Detailed studies of two fields in Nebraska reported by Gerhards *et al.*, (1997) also demonstrated that weeds were patchy, and that the weed patches were stable, with seedling distributions in one year being a good predictor of future distributions. A more extensive study reported by Lutman *et al.* (1998) also provides strong information to show that grass weeds in cereal crops often have a patchy distribution.
2. The costs of treating grass weeds in a grass crop with selective herbicides is relatively high, and studies reported by Lutman *et al.* (1998) and Rew *et al.* (1996*a*) indicate that savings in herbicide costs could be in the range of £5–12 ha/year, depending upon weed distribution and the treatment strategy applied.
3. There is considerable environmental pressure to reduce herbicide use, particularly for some herbicides that are now being found in groundwater and river systems at levels that require treatment if they are to be used as a source of drinking water.

Spatially variable applications of herbicide can be made in fallow or widely spaced row-crops, such as maize or soya beans, by using spectral reflectance type detectors to determine the presence of weeds, and to actuate a spray application system directly. This approach has been developed commercially both in Australia (Felton, 1995), and in the USA, and considerable savings in herbicide use have been demonstrated – see also Chapter 3. The same approach has recently been developed in Europe for use in amenity areas, where the presence of weeds in pavements and gravel paths can be detected by systems working on spectral characteristic criteria.

While the approach of detecting weeds automatically using the spectral reflectance characteristics of light from leaf and soil surfaces works well when weeds are growing against a non-vegetative background, the method is much less successful when weeds are in a growing crop, and particularly in a grass crop. For such situations, an approach based on the use of a control map has been proposed (Paice *et al.*, 1995; Lutman *et al.* 1998). The use of a map has advantages relating to:

- the ability to detect weeds when detection may be easier, rather than at the time of spraying;
- the transformation of a weed patch map into a treatment map, taking account of the dose-response characteristics of the selected appropriate tank mix; the likely changes in the weed patch positions between detection and treatment, relating to field operations such as harvesting or cultivations (Rew and Cussans, 1997); uncertainty in the performance of field location systems both when generating the weed patch map and when applying the treatments; and the operating characteristics of the patch sprayer.

Strategies used in the transformation of the weed to treatment map can take account of environmental as well as economic factors. For example, in areas of the field where the weed pressure has been assessed to be low, it may be appropriate to use a herbicide that is relatively environmentally benign but with a limited efficacy, whereas in areas assessed to be at high risk of weed competition, a more efficacious but more environmentally damaging herbicide could be used.

A range of strategies for treating weed patches is possible. The simplest is on/off control based on the application of a given tank mix loaded into the sprayer. This can be readily implemented using a conventional boom sprayer, but must be based on a high level of confidence in the weed detection system if substantial savings are to be made and weed control maintained over a number of cropping seasons. The application of multiple dose levels to different parts of a field can again be based on the application of a predetermined tank mix with low insurance dose levels applied to areas of a field assessed as having low weed pressures, and higher doses used where weed pressure poses a higher threat to yield, quality and ease of harvesting. Work by Paice *et al.* (1997) used a computer simulation model to demonstrate that the use of a dual-dose spatially variable treatment strategy gave savings in herbicide use that could be sustained over a ten-year period whereas, with typical weed patch sensing performance, on/off control gave weed populations that tended to increase over a ten-year period and involved increasing levels of herbicide use. The use of injection metering enables the mixtures of components of a tank mix to be varied across a field, as well as dose levels. This then enables strategies developed on either financial or environmental terms to be implemented as discussed above.

The use of a patch spraying system based on a treatment map requires:

- a means of in-field location for both weed patch detection and application control;
- a means of detecting weed patches that would ideally recognise weed density and species as well as location;
- methods of generating and supporting a treatment map on the application system; and,
- a system for controlling sprayer output.

A number of systems have been developed for determining in-field location, including the use of tramlines, radio beacons and laser-based range finders. In recent years, satellite navigation systems, particularly the Global Positioning System (GPS) operating in differential mode to give typical location accuracies of ±2.0 m for 95% of the time, have been developed for precision farming systems. Such systems are now widely available commercially, and are likely to form the basis for the development of spatially variable application systems in the next decade.

As indicated above, automatic detection of green weeds against a soil background is technically feasible and has been developed commercially. While the automatic detection of weeds in crops is the subject of much current research effort using combinations of image and spectral reflectance analysis, to date no robust commercial systems have been developed. Weed patch detection must therefore be based on human recognition. Methods to enable the results from manual 'scouting' have been developed, e.g. Stafford *et al.* (1996); for use in field walking, from survey vehicles and those conducting other field operations such as fertiliser spreading and combine harvesting. Initial developments used key pads or push-button panels to record weed patch positions, species and densities, but practical experience showed that these were often tiring and difficult to use. Recent developments have shown that voice recognition computer systems can be used in a range of practical weed patch assessment situations, with the advantage of speed and ease of use. It is recognised that weed patch mapping in the field, based on manual discrimination between weed and crop, is time-consuming and costly. It is probably the major factor limiting the uptake of such systems, and pressure to develop automatic systems is expected to increase. An example of a typical weed map produced by detailed manual surveying is shown in Figure 4.5.

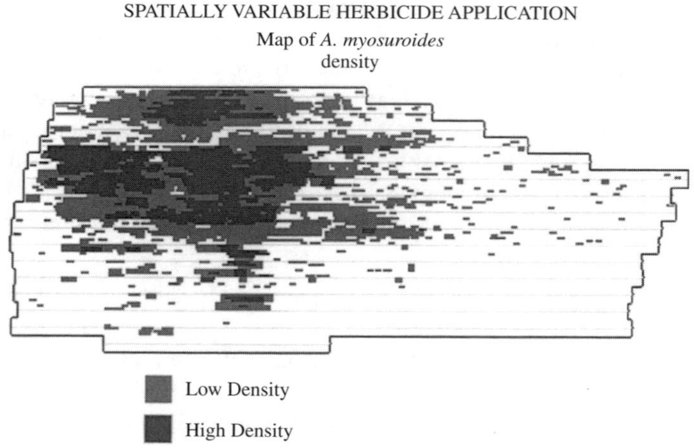

Figure 4.5 An example of a weed patch map generated by detailed manual survey

The commercial development of computer-based control systems for tractors and other agricultural implements has provided a platform that enables coded mapped information to be effectively transferred from a desktop computer to the tractor cab. Initially developed in conjunction with GPS field location for yield-mapping applications, this type of system is now being incorporated into tractors and specialised application vehicles (see Figure 4.6). These can be used to give access to the treatment map in the field, and hence control signals for the application system. Data transfer is commonly by 'smart cards' with current systems, but it is likely that future developments will use radio/telephone connections to automate the data transfer process (Miller, 1999).

Methods by which dose and mixtures of spray chemicals can be delivered at different controlled rates have been discussed earlier in this section. For spatially variable (patch) applications, Miller *et al.* (1997) summarised the performance requirements of the application system as follows:

- accuracy of delivered dose to better than ±5% of target;
- spatial resolution to be in the range 4.0 to 6.0 m;
- response time between delivered dose levels to be less than 1.0 s;
- delivery to target areas over a dose rate range of at least 5 : 1;
  plus the ability to work with a wide range of formulations.

The question of resolution of treatment has been considered in a number of studies. Work by Rew *et al.* (1996*a*, 1997) showed that it was likely that the

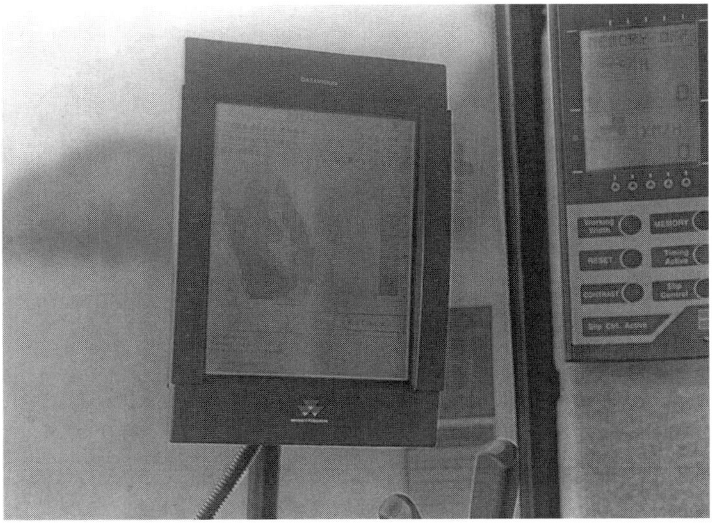

Figure 4.6 Typical computer-based control system for use with agricultural equipment capable of supporting digital treatment map information and with a GPS interface

largest savings in herbicide use would come from using a spatial resolution of 4.0 to 6.0 m when treating grass weeds in cereal crops. This assessment was based on a study of a large number of weed maps collected under typical farming conditions, but probably did not account fully for the increased costs of field mapping when operating at smaller spatial scales.

When developing a commercial patch spraying system, the work described by Miller *et al.* (1997) considered options based on injection metering and concluded that these were too expensive for first-level patch spraying operators seeking to retrofit systems to existing machines. A system was therefore developed based on the use of conventional nozzles, with typically three nozzles being used at each nozzle position on the boom. The sprayer control system is programmed with nozzle flow rates and spray qualities at different pressures, such that signals from the treatment map controller relating to required dose and forward speed can be translated into a defined nozzle operating at a predetermined pressure. This then enables a conventional nozzle to be used to deliver a range of flow rates over a range of more than 5 : 1 at a given spray quality. Figure 4.7 shows this system operating in the field.

Much of the field trials work with spatially variable (patch) spraying systems has used herbicides, and has shown that typically some 50% of the chemical needed for grass weed control in cereal crops can be saved. Other weed species are also known to be patchy (cleavers and thistles), and therefore overall estimated savings in herbicide use are commonly around 25%. Research work is currently examining the scope for the spatially variable application of

Figure 4.7 Spatially variable spraying system operating under typical field conditions

other plant-protection products, particularly fungicides and plant-growth regulators. It is therefore likely that in the future, as practical systems continue to be developed at increasingly competitive costs, ways will be found of using these technologies, both within conventional and developing agronomic practices. The combined drivers of cost saving and reduced environmental loading, with environmental safety requirements will ensure that such systems are developed and used.

REFERENCES

Anon. (1986) *An Evaluation of Safety Features on Agricultural Crop Sprayers*. Report by Health and Safety Executive, HM Agricultural Inspectorate, Nottingham.

Anon. (1996*a*) *British Standard BS 6365: Spraying Equipment for Crop Protection. Part 8, Specification for Induction Hoppers*.

Anon. (1996*b*) *British Standard BS 6356: Spraying Equipment for Crop Protection. Part 9: Specification for Systems for Closed Transfer of Liquid Formulations*.

Antuniassi, U.R., Miller, P.C.H. and Paice, M.E.R. (1997) Dynamic and steady-state dose responses of some chemical injection metering systems. *Proceedings, Brighton Crop Protection Conference – Weeds*, pp. 587–692.

Brazelton, R.W., Akesson, N.B., Maddy, K.T and Yates, W.E. (1981) *Progress in Pesticide Worker Safety in California*. American Society of Agricultural Engineers, Paper No. 81–5001.

Cooper, S.E. and Taylor, W.A. (1998) A survey of spray operators' agrochemical container rinsing skills conducted in June 1997. *Proceedings, BCPC Symposium No. 70: Managing Pesticide Waste and Packaging*, pp. 145–148.

Felber, H.U. (1993) Closed transfer systems for small volume refillable containers. *Proceedings, ANPP–BCPC Second International Symposium on Pesticide Application Techniques*, Strasbourg, **II**, 479–486.

Felton, W.L. (1995) Commercial progress in spot spraying weeds. *Proceedings of British Crop Protection Conference – Weeds*, 1087–1098.

Frost, A.R. (1990) A pesticide injection metering system for use on agricultural spraying machines. *Journal of Agricultural Engineering Research*, **46**, 55–70

Frost, A.R. and Miller, P.C.H. (1988) Closed chemical transfer systems. In: *Weed Control in Cereals and the Impact of Legislation on Pesticide Application*, Aspects of Applied Biology, 18, 345–359.

Garnett, R.H. (1993) Closed chemical transfer systems; objectives and definitions. *Proceedings, ANPP–BCPC Second International Symposium on Pesticide Application Techniques*, Strasbourg, **II**, 471–478.

Gerhards, R., Wyse-Pester, D.Y., Mortensen, D. and Johnson, G.A. (1997) Characterising spatial stability of weed populations using interpolated maps. *Weed Science*, **45**, 108–119.

Gilbert, A.J. (1998) Design guidelines, features and performance characteristics and development of current pesticide containers. *Proceedings, BCPC Symposium No. 70: Managing Pesticide Waste and Packaging*, pp. 9–16.

Giles, D.K. and Comino, J.A. (1990) Droplet size and spray pattern characteristics of an electronic flow controller for spray nozzles. *Journal of Agricultural Engineering Research*, **47**, 249–269.

Goldsworthy, P.E. and Carter, P. (1998) The safe disposal of clean agrochemical containers on farm – an interim report. *BCPC Symposium No. 70: Managing Pesticide Waste and Packaging*, pp. 85–88.

Gussin, E. (1998) Developing the Ecomatic system. *Proceedings, BCPC Symposium No. 70: Managing Pesticide Waste and Packaging*, pp. 141–144.

Hancock, T. (1993) Plastic containers for agrochemicals into the 1990s. *Proceedings, ANPP – BCPC Second International Symposium on Pesticide Application Techniques*, Strasbourg, pp. 1–8.

Hibbitt, C.J. (1998) Packaging waste management – a key issue for the UK agrochemical industry. *Proceedings, BCPC Symposium No. 70: Managing Pesticide Waste and Packaging*, pp. 3–7.

Landers, A.J. (1993) Direct injection sprayers – a method of reducing environmental pollution. *Proceedings, ANPP – BCPC Second International Symposium on Pesticide Application Techniques*, Strasbourg, pp. 305–312.

Lavers, A. (1993) An investigation into the efficiency of two methods of rinsing empty crop protection chemical containers. *Proceedings, ANPP–BCPC Second International Symposium on Pesticide Application Techniques*, Strasbourg, **1**, pp. 15–24.

Lutman, P.J.W., Rew, L.J., Cussans, G.W., Miller, P.C.H., Paice, M.E.R. and Stafford, J.V. (1998) Development of a 'patch spraying' system to control weeds in winter wheat. *HGCA Project Report No. 158*. Home-Grown Cereals Authority, London.

Miller, P.C.H. (1999) *Automatic recording by application machinery of rates and spatial distribution of field inputs. Proceedings No. 439*. The International Fertiliser Society, York.

Miller, P.C.H., Paice, M.E.R. and Ganderton, A. (1997) Methods of controlling sprayer output for spatially variable herbicide application. *Proceedings, Brighton Crop Protection Conference – Weeds*, 641–644.

Miller, P.C.H. and Stafford, J.V. (1991) Herbicide application to targeted patches. *Proceedings, Brighton Crop Protection Conference – Weeds*, 18–21, 1249–1256.

Miller, P.C.H., Tuck, C.R., Gilbert, A.J. and Bell, G.J. (1991) The performance characteristics of a twin fluid nozzle sprayer. *Proceedings, Air Assisted Spraying in Crop Protection, 7–9 January 1991*. British Crop Protection Monograph, **46**, 97–106.

Paice, M.E.R., Miller, P.C.H. and Bodle, J.D. (1995) An experimental sprayer for the spatially selective application of herbicides. *Journal of Agricultural Engineering Research*, **60**, 107–116.

Paice, M.E.R., Miller, P.C.H. and Day, W. (1977) Using a computer simulation to compare patch spraying strategies. *Proceedings, 1st European Conference on Precision Agriculture*, 421–429.

Paice, M.E.R., Miller, P.C.H. and Day, W. (1996) Control requirements for spatially selective herbicide sprayers. *Computers and Electronics in Agriculture*, **14**, 163–177.

Paice, M.E.R., Miller, P.C.H. and Power, J.D. (1993) A practical pesticide injection metering system for agricultural sprayers. *Proceedings, ANPP–BCPC Second International Symposium on Pesticide Application Techniques*, Strasbourg, **1**, 313–320.

Perryman, M.A. (1993) The development and use of coupling system for small volume refillable containers. *Proceedings, ANPP–BCPC Second International Symposium on Pesticide Application Techniques*, Strasbourg, **II**, 487–494.

Rew, L.J. and Cussans, G.W. (1997) Horizontal movement of seeds following tine and plough cultivation: implications for spatial dynamics of weed infestations. *Weed Research*, **37**, 247–256.

Rew, L.J., Cussans, G.W. and Miller, P.C.H. (1996*b*) Evaluation of four distance and navigation methods for mapping weed positions within arable fields. *Proceedings, 2nd International Weed Control Congress, Copenhagen*, 1103–1108.

Rew, L.J., Miller, P.C.H. and Paice, M.E.R. (1997) The importance of patch mapping resolution for sprayer control. *Aspects of Applied Biology*, **48**, *Optimising Pesticide Applications*, 49–55.

Rew, L.J., Cussans, G.W., Mugglestone, M.A. and Miller, P.C.H. (1996*a*) A technique for surveying spatial distribution of *Elymus repens* L. and *Cirsium arvense* L. in cereal fields and estimates of the potential reduction in herbicide use from patch spraying. *Weed Research*, **36**, 283–292.

Rietz, S., Pályi, B., Gunzelmeier, H., László, A. (1997) Performance of electronic controls for field sprayers. *Journal of Agricultural Engineering Research*, **68**, 399–407.

Smith, R.K. (1998) Effective container cleaning for crop protection products. *Proceedings, BCPC Symposium*, Strasbourg, *No. 70: Managing Pesticide Waste and Packaging*, 71–84.

Stafford, J.V., Le Bars, J.M. and Ambler, B. (1996) A hand-held data logger with integral GPs for producing weed maps by field walking. *Computers and Electronics in Agriculture*, **14**, 101–120.

Wilson, B.J. and Brain, P. (1991) Long term stability of *Alopecurus myosuroides* Huds. within cereal fields. *Weed Research*, **31**, 367–373.

5 Specialised Application Technology

G.A. MATTHEWS
Imperial College of Science, Technology and Medicine, Silwood Park, Ascot, Berkshire, SL5 7PY, UK

INTRODUCTION

In addition to the large-scale tractor-mounted and trailed or self-propelled applicators used in arable and orchard spraying of pesticides, described in the two previous chapters, a vast range of other types of equipment, including manually carried and trolley-mounted sprayers, is used on smaller areas, for example, by local authorities and others to treat amenity areas, sports grounds, public health pests and in other specialised areas, including glasshouse and forest management. Localised treatment may be an essential component of IPM programmes to ensure reduction of pest populations without decimating natural control agents. Aerial application also continues to be important on very extensive areas of farmland and forests and, in certain circumstances, for large-scale control of vectors of human diseases, such as dengue haemorrhagic fever. This chapter, as the following chapter, considers how these specialised forms of pesticide delivery can be optimised to reduce costs and environmental contamination, and also improve operator safety.

GROUND EQUIPMENT – HYDRAULIC SPRAYERS

The main types of equipment considered here are the hydraulic sprayers, such as compression (Figure 5.1) and lever-operated knapsack sprayers (Figures 5.2, 5.3) normally used for residual applications to surfaces. Hydraulic sprays may also be applied with motorised pumps on knapsack sprayers, or larger units are carried on a trolley (Figure 5.4) or small vehicle. Pesticides formulated for dilution in water are frequently applied at volumes of around 100–500 litres per hectare. Other types of space treatments and more specialised equipment are considered later.

Apart from small motorised hydraulic sprayers, similar to the tractor-mounted equipment but on a miniature scale, the majority of small ground-based sprayers are manually carried or, if trolley-mounted with a hose and lance, are manually operated. A common feature of these sprayers is that the actual delivery of a

Optimising Pesticide Use Edited by M. Wilson
 ISBN: 0-471-49075-X

Figure 5.1 Compression sprayer being used to apply a residual insecticide spray to control mosquitoes in Africa. Photo. G. A. Matthews

pesticide is entirely dependent on the skill and consistency of work achieved by the operator. Application rate is affected by the output of the nozzle, the swath treated and speed of travel.

Nozzle output is influenced by operating pressure, so, with manually operated pumps on basic lever-operated knapsack sprayers, the pressure will vary

Figure 5.2 Lever operated knapsack sprayer. Photo. Cooper Pegler Division, Hardi.

depending on the speed of pumping, while, on compression sprayers, the pressure falls as the tank is emptied. Adjustable pressure regulators have been available for a long time, but few users have fitted them due to the extra cost, or, if originally fitted, they may be removed due to blockages when pesticide deposits dry inside the mechanism. The actual pressure at the nozzle will also depend on

Figure 5.3 Sprayer with a shield around nozzle to reduce drift. Photo Micron.

how the valve is set and needs to be checked with an accurate pressure guage. More recently, an adaptation of the diaphragm check valve has been developed to provide a constant pressure at the nozzle. Colour-coded versions of this valve are designed for use at 1, 1.5, 2 or 3 bar pressure, the pressure being selected primarily in terms of the pesticide used, thus the 1 bar valve is for herbicide use to minimise risk of spray drift (McAuliffe, 1999) (Figure 5.5). These valves also must be cleaned after use by pumping water through the sprayer, but, by

Figure 5.4 Motorised sprayer on trolley for use in glasshouse. Photo Hardi.

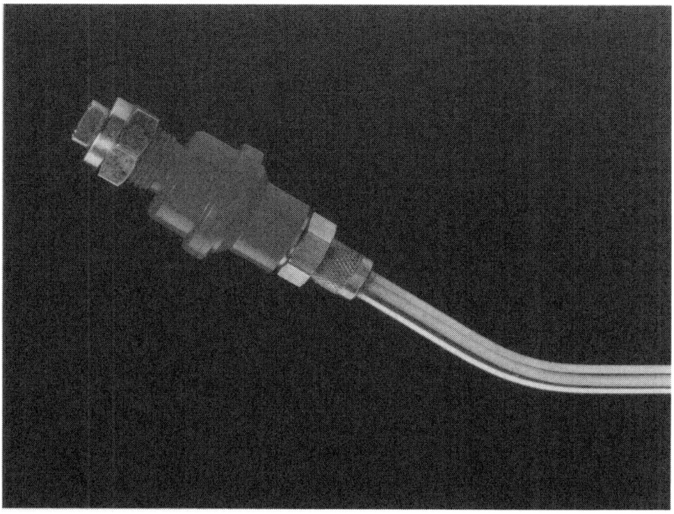

Figure 5.5 Control Flow valve (CFV) on a lance. Photo GATE llc.

regulating nozzle output, using this type of valve eliminates one of the main causes of variation in dosage applied.

Swath width is basically a function of the spray angle, provided that the user maintains the nozzle at a constant distance from the target area being treated. Some users have added a weighted chain or string to part of the lance as a guide to maintain holding a lance at a set height while applying herbicides downwards to the ground. Often, however, when treating a larger area of foliage, a nozzle is waved over the weed or crop. Under these circumstances, swath width is more likely to be a function of the width of the inter-row or pathway, and walking speed may be reduced. This may result in an excessive spray volume being applied, and much of the spray is wasted as it drips from the foliage and contaminates the soil. Lower-output nozzles can be used, but they have been ignored where particles suspended in poor-quality water have caused blockages, although efficient filters within the sprayer would eliminate this problem.

Walking speed will vary, depending on a number of factors. The stature, fitness and training of the operator, the type of equipment (e.g. weight, comfort), requirements for personal protective clothing (e.g. face mask, respirator) the terrain (e.g. hillside, flat), climatic conditions (e.g. hot, humid) and mode of spraying (e.g. crop target, pesticide) are some of the factors. On reasonably level ground in a row crop, a speed of 1 m/s can usually be maintained for long periods, but, where the foliage from adjacent rows meets and restricts the movement of the operator, the speed will be slower. Some equipment incorporates an aural indicator, usually set to bleep every second as a guide to walking speed.

In view of these variations, the calibration of equipment should be done by individual users under the conditions in which they will apply the pesticide. Measurement of each of the three factors (nozzle output, swath and speed) can be measured and the volume application rate calculated using the formula:

$$\frac{\text{Nozzle output (litres/minute)}}{\text{swath (m)} \times \text{walking speed (m/minute)}} = \text{litres/m}^2 \times 10\,000 = \text{litres/hectare}$$

A test using this system is part of the Certification for spray operators in the United Kingdom.

An alternative is to fit a calibrated container, referred to as a 'Kalibottle' (Figure 5.6), to the nozzle and to collect the spray while treating 25 m^2. The volume application rate is then read from the side of the bottle. This technique has the advantage when training teams of spray operators, as they can see the effects of walking speed or other changes in application technique.

Selection of the nozzle and accurate calibration are crucial if the optimal dosage is to be applied. The tendency in the past has often been to recommend a dose per hectolitre and assume that the operator will apply at least 1000 litres per hectare, using a nozzle with a large orifice. Whereas this may be possible in areas with a good water supply and motorised pumping system, when such a recommendation is translated into manual equipment, most operators will use fewer tankloads of spray and apply rather less than 1000 litres, so unless the spray concentration is increased, less active ingredient will be applied. In practice, the lower dosage will be an advantage where a high proportion of the pesticide would have been lost due to 'runoff' with a high volume treatment. Simply putting on fewer loads

Figure 5.6 Training spray operators to calibrate a sprayer using a 'Kalibottle'. Photo G. A. Matthews

with an inappropriate nozzle may control a pest, but to optimise application of a minimal amount of pesticide requires adequate coverage where the pest is located. The choice of a fan, cone or deflector nozzle and fine, medium or coarse spray quality discussed in Chapter 3 is similar with manually operated equipment. With the need to minimise the weight of spray carried, preference (where possible) is given to low-output nozzles, but careful filtration is then important. Deflector nozzles have been used more on manual sprayers for weed control, as the orifice is less likely to block.

When the volume rate has been determined, it is then possible to calculate the appropriate quantity of the pesticide product to put into each tankload. Where small manually carried sprayers are used, this may require measurement of very small quantities. Often a container cap is regarded as the measure, but this can lead to spillage and contamination of fingers unless the user is wearing protective gloves, although containers with a built-in measure (Figure 5.7) are now available. Equipment to facilitate transfer of measured doses from large containers to knapsack sprayers is also available. However, changes in formulation have now produced some pesticides in tablet form, while others are marketed as dispersible granules in water-soluble sachets, so that one of these tablets or sachets is added to each tankload. So far there is no commercial direct injection system for knapsack sprayers, although one design was described by Craig *et al.* (1993) in which the pesticide was held in a flexible bag within a container. Water at constant pressure was fed into the container, outside the bag, and pesticide injected through a dip tube and metering system into the water entering the lance. The container was designed to be reused, as the internal bag was not rinsed.

The versatility of hydraulic sprayers enables them to be used in a wide variety of situations. There are BSI standards for the compression and lever-operated knapsack sprayers. The relatively inexpensive compression sprayers have a small air pump to pressurise a container of 0.5–10 litre capacity, which is filled with the spray liquid to about three-quarter capacity. Pumping is not required while

Figure 5.7 Measuring dose of insecticide using a container with built-in measure. Photo G. A. Matthews

spraying, so the user can give attention to the movement of the nozzle. Plastic compression sprayers are widely available for gardeners. Stainless-steel sprayers have been used in public health work to treat the wall surfaces of dwellings with a fan nozzle for mosquito adulticiding, and, with other nozzles, for larviciding small areas of water. Similar sprayers are used to treat hotels, restaurants, other buildings and aircraft for cockroach control, either using a cone nozzle or, in some situations, a straight jet to penetrate cracks and crevices in which insect pests may hide. The aim is to restrict the treatment to localities, such as under work-benches, sinks and behind panelling, where the pesticide will have the most impact and avoid surfaces where deposits will be soon washed off.

In agricultural and horticulture, preference is generally given to the lever-operated knapsack sprayer fitted with a diaphragm or piston pump which has to be operated continuously. Invariably, these sprayers are fitted with a single lance and nozzle, which is held in front of the operator due to the perceived need to see the spray. This practice unfortunately always results in operator contamination, and especially in field and bush crops when the crop canopy is high. Ideally, the nozzle should always be used downwind of the operator, and there have been designs for rear-mounted nozzles to reduce operator contamination, but these are considered too heavy by users. Multiple nozzles mounted on a boom can be used if the pump capacity is sufficient. To reduce the drudgery of manual pumping, similar sprayers are available, fitted with a rechargeable battery and electrically operated pump or a two-stroke engine and pump. Some sprayers are fitted with a pressure limiter, but this is not as effective as having a pressure-control valve.

An important feature of using these manual sprayers is that it is possible to do selective spot treatments, for example, confining the application of herbicides to patches of weeds. Also, where there is concern about downwind spray drift, placement of the nozzle nearer to the target and using a low pressure can enable treatments closer to a sensitive area. In conjunction with an attractant such as molasses or protein hydrolysate, insecticide bait sprays have been used in discrete spots for fruitfly control. Knapsack sprayers are often used in difficult terrain where access to vehicles is difficult or impossible. However, the lack of water in some situations, and the high requirement for labour, have created a demand for alternative application systems (see below).

In larger areas, small sprayers are mounted on vehicles, such as a four-wheel drive Land Rover™, but when treating open areas of close-mown grass, spray is vulnerable to gusts of wind. This can be reduced by adding a perforated shroud to allow airflow while baffling stronger winds. Inside the semi-circular shroud, the spray bar is angled 60° forward (Turner, 1996; Matthews, 1999).

SPACE TREATMENTS

In enclosed areas such as glasshouses and in storage sheds, a pest problem may require the whole area to be treated rapidly. Similarly, rapid treatment of disease vectors, especially mosquito vectors of malaria and dengue, may involve space treatments both inside dwellings and outside in urban residential areas. Space treatments are predominantly with insecticides, but fungicides have been used in horticulture. An unusual use of a herbicide is to suppress sprouting of potatoes in store. Space treatments are carried out with extremely small droplets, with a VMD $< 50\,\mu m$ and generally referred to as fogs, that remain airborne as long as possible. Fogs are generated by two distinct methods: the thermal method or by using an airshear nozzle (vortex) without heat (known as cold fogging). Thermal fogs are preferred by some users due to the very opaque dense white fog that is produced (Figure 5.8) but the pesticide is exposed very briefly to a high temperature (*c*.500°C), so some degradation of the pesticide may occur.

One advantage of the cold fog is that the amount of diluent is minimal and ultra-low volume applications are possible, compared with a thermal fog where the pesticide is diluted in a large volume of carrier. The diluent was usually based on oil or diesel, but, rather than use specialised formulations, water is increasingly used with conventional formulations in combination with an adjuvant. Secondly, cold fog treatments can be carried out when the area is unoccupied, by using a remote control/timer, while constant supervision of a thermal fogger is necessary.

In contrast to the residual spray treatment, the dosage of pesticide applied as a fog is low when flying insects are to be killed. There is generally no intention of providing a residual deposit as the pesticide is deposited virtually only on the top of leaves and other relatively horizontal surfaces. A sequential series of treatments is used if reinfestation occurs. However, the technique has been

Figure 5.8 Space treatment in a plantation crop, using a thermal fogger. Photo G. A. Matthews

used to apply fungicides and some insecticides, such as *Bacillus thuringiensis*, where the action of foliar deposits is crucial. Normally, treatments are done in the evening to avoid intense sunlight and allow sufficient time after treatment for the small droplets to sediment within the treated area. The alternative technique for foliar deposition is to use a mist, defined as a very fine spray with a VMD between 50 and 100 μm, and less than 5% of the spray with droplets <30 μm diameter. The important distinction between a mist and a fog is that the risk of inhalation is less and that the larger droplets are collected more rapidly on the foliage. In practice, some care is needed, as certain manufacturers have described cold fogging equipment as a low-volume mister. Equipment to produce mists is discussed below, under CDA and motorised mistblowers.

The dilemma in many horticultural pest problems is how to satisfy the commercial requirement for blemish-free produce, while at the same time ensuring that there are preferably no residues of any pesticide. Frequent harvesting over several months also demands that labourers are in close contact with treated crops. While low dosages are applied by space treatments, care is needed when applications are applied too frequently. Furthermore, the ventilation required before re-entry of an area treated with a fog or mist will dissipate any pesticide still airborne to other areas outside, although the amount airborne after six hours is relatively small.

In some situations, foliar treatments can be avoided by using a prophylactic soil-applied pesticide. Treatment of potted plants, especially ornamentals, may be by an admixture of granules to the potting compost, by dipping the roots into a slurry of pesticide at planting, or by a subsequent drench. However, all these treatments have a limited persistence, and are only aimed at protection during the first two to six weeks of plant development.

OTHER GROUND EQUIPMENT

WEED WIPERS

Weed wipers originated in the USA, when clumps of tall grasses in fields needed to be controlled without the herbicide contacting the crop. Initial efforts used a series of horizontally projected jets across the top of the crop which, if the weed was not intercepted, were collected and recycled. Splashes from the weeds could affect the crop, so the idea of smearing the herbicide on the weeds from an absorbent pad was tried (Figure 5.9). The equipment is designed to exploit the height differential between the weeds to be controlled, such as nettles, thistles, and other vegetation. The most widely used herbicide is glyphosate, as it is translocated down from the upper foliage, provided that the weed is not affected by drought. Only herbicides approved for use in weed wipers may be used.

The main problem with the development of weed wipers has been to control delivery of sufficient liquid to the pad, so that it is wet enough to transfer the

Figure 5.9 Selective application of a herbicide with a weed wiper. Photo G. A. Matthews

herbicide to the weeds without dripping. If the pad is too dry, inadequate transfer occurs. To overcome these problems, the herbicide is mixed with a polymer to increase its viscosity, which can be checked by timing the flow through a special viscosity cup. However, with increased viscosity it may take time for the whole wiper to be covered with herbicide.

There are a number of different designs of weed wipers, which range in size from the hand-carried 'hockey-stick' to large units mounted on tractors. In the latter case, a series of wipers can be mounted to move independently in relation to the contours of the field. Choice of applicator will depend on the area to be treated and the frequency of treatment. As the amount of foliage treated is selected, the total volume of herbicide applied is much less than if a whole area was sprayed. In the UK, apart from use on beet crops, the main emphasis has been to control weeds in 'set-aside' land and in SSSIs (subject to approval from English Nature's local staff and requirements of the Wildlife and Countryside Act 1981), as the problem of spray drift is eliminated, although with some herbicides there remains the possibility of vapour drift. Care is needed where ragwort occurs, as the dried weed remains a danger to livestock.

A disadvantage of a weed wiper is that when a treatment is complete there can be a large area of the wiper still contaminated with herbicide. This has to be hosed off with water, using a brush to ensure penetration of the absorbent surface. It is important to ensure that the herbicide contaminated washings can be disposed of safely and, when dry, the wiper should be covered by a protective sleeve.

CDA SPRAYERS

One of the problems with manually carried sprayers is the weight of water that has to be carried. Maximum loads that can be lifted under European legislation limit the weight of knapsack equipment to 25 kg. In the semi-arid areas where water is scarce, equipment with rotary atomisers had already been designed for ultra-low volume oil-based formulations, and this has now evolved into very low-volume application using up to 20 litres of water per hectare. With conventional hydraulic nozzles producing a wide range of droplet sizes, much of the pesticide is wasted in the largest droplets, while the smallest lead to spray drift. The need for a narrow droplet spectrum led to the concept of controlled droplet application (CDA), for which the rotary atomiser was much more suitable. Initially, equipment with spinning discs applied insecticides in droplets with a VMD of about 70–100 μm, (Figure 5.10) and it was realised that, in order to minimise downwind drift of herbicides, a larger droplet size was required (Figure 5.11). However, certain herbicides have been used with small droplets, when controlling extensive areas of particular weeds such as bracken and heather. Studies indicated that, for most weeds, a compromise of 250 μm was needed to minimise drift, yet give adequate coverage of the weeds. Downwind displacement of such large droplets is minimal, especially as they are usually released within a few centimetres of the ground. Even larger droplets have been used with certain herbicides, such as glyphosate, as this chemical is readily translocated by the weed.

As minimal volumes are needed with these sprayers, the original idea with ULV sprays was for manufacturers to supply the pesticide in pre-packaged containers to eliminate the need for the user to prepare sprays from a concentrated formulation. A range of ULV pesticides was developed, especially for glasshouse use, with a spinning disc mounted in front of a battery driven, mains electric, or two-stroke engine driven fan (Figure 5.12). Such equipment is also used for fly control in animal houses, and for the vaccination of chickens. The main advantage of the system is that the concentration of the spray is fixed by the chemical manufacturer and, by using appropriate relatively involatile formulations, droplet size did not decrease significantly in flight between the nozzle and deposition on the target. The pre-pack concept was also marketed with the 'Electrodyn' electrostatic sprayer, where the 'Bozzle' container incorporated a nozzle (Coffee, 1981). Other types of pre-packed container have been used with rotary atomisers for weed control, particularly where pavements and other areas require treatment in urban areas, as they allow the operator to treat a large area each day, without the need to transport water. Alternatively, the user can carry some of the spray in a small knapsack container.

MOTORISED KNAPSACK MISTBLOWERS

Airshear nozzles are used on motorised mistblowers with a centrifugal fan that creates a high-velocity airstream. Users perceive that the ability to project the spray over several metres is an advantage, and air disturbance within the crop canopy is expected to improve the spray distribution. However, successful application with this equipment depends very much on how they are operated. The air velocity should not be adjusted, as the engine should always be used at the

Figure 5.10 CDA rotary atomiser insecticide sprayer. Photo Micron Sprayers

optimum speed. If the engine is throttled back, apart from increased wear on the engine, the shearing over the spray liquid into droplets is inadequate, and larger droplets are produced. Control of liquid flow is possible, but generally the lowest flow rate should be chosen, as this produces the smallest 'mist' spray that will remain projected in the airstream. Large droplets tend to fall out of the airstream very rapidly, so, if the user is endeavouring to project spray upwards

Figure 5.11 CDA rotary atomiser herbicide sprayer. Photo Micron Sprayers

into tree canopies, a high proportion of the spray volume will be deposited on the ground under the trees. Similarly, users, exploiting the airstream to achieve a wide swath horizontally over a crop, will find that the larger droplets will give a peak deposit close to the nozzle. Although air velocity is dissipated very rapidly, it is also unwise to have foliage within the first 1–1.5 metres from the nozzle, as the position of leaves will be influenced by the air velocity, and deposition of spray may be adversely affected. Even with the high air velocity, spray should not be directed upwind, as this will be blown back on the operator. Furthermore, if the spray is directed above a crop canopy, even though spray may appear to cover a large number of rows, the smallest droplets may be lost in thermal upcurrents of air, and penetration of the canopy may be inadequate.

Compared to spraying with hydraulic nozzles, the volume applied is generally lower and the spray concentration increased in proportion to apply the same dosage per hectare. Care is needed to ensure that the higher concentration is not phytotoxic to foliage. The droplet spectrum will depend on the actual design of the airshear nozzle, and is better on equipment in which the liquid is spread more thinly before the air forms the spray. On some mistblowers, a rotary nozzle is used to improve the droplet spectrum, and, in one new design intended initially for coffee spraying, the rotary nozzle is mounted with a propeller fan behind the operator.

SEED TREATMENTS

Confining a small dosage of pesticide to a seed is an ideal selective application technique. Seedlings can be protected at a critical stage when an insecticide spray

Figure 5.12 Mist spray with air-assisted CDA sprayer in a glasshouse. Photo Micron Sprayers

application would be inefficient and would be likely to have an adverse effect on natural enemies. Care is needed in order to have sufficient active ingredient adhering to the surface of the seed to provide sufficient protection, without any adverse effect on seeds during storage and subsequent germination. Furthermore, it is essential that the amount of active ingredient is consistent and that some seeds are not overdosed. Losses of active ingredient from the seed during sowing must also be avoided, as blockages of the seed drills may occur. These aspects are to a large extent overcome by seed pelleting, which allows irregular small seeds to be built up into spherical capsules, that facilitate precision planting. One or more active ingredients are mixed with materials such as cellulose powder and adhesive materials (e.g. polyvinyl polymers and starch) to form a thicker coating around individual seeds. Nutrients may also be added to the pellet to ensure robust establishment of a crop. Discrete layers of the pellet can be built up to separate different materials (Halmer, 1988), using different coating techniques (Jeffs and Tuppen, 1986; Clarke, 1988; Clayton, 1993).

GRAIN TREATMENT

In contrast to seed treatment, protection of grain and other produce requires handling large quantities very quickly. Reference is made elsewhere (Chapter 9 of this book) to modified atmospheres as an alternative to fumigation. However, such techniques require airtight facilities, but, in many parts of the world, grain is stacked in sacks in the open or ventilated warehouses. Admixture of grain is possible, especially if incoming grain is known to be infested with insects, but care is needed to avoid unacceptable residues and inhalation of pesticide and dust particles during the grain treatment process. Admixture is possible by fixing a nozzle, such as an even-spray fan, above a conveyor belt, so that incoming grain is sprayed as it enters a store. Preference may be given to fumigation under sheets and subsequent sequential space treatments to minimise the risk of reinfestation, combined with a residual treatment of the store cleaned thoroughly prior to the arrival of the grain.

APPLICATION OF DRY PARTICULATES (DUSTS, GRANULES, BAITS)

While the majority of pesticides are applied as sprays, there are a number of situations where the active ingredient is applied with various inert materials as a dry particulate. Dusts with particles generally below 10 μm were more widely used, but, as these particles are so prone to downwind drift, their use is now very much restricted. Sulfur dust is still used in some circumstances, especially where a spray will increase humidity, as in some polytunnels. Granulating dusts into larger drift-free granules has now become the main method of applying dry particles. The modern technology of coating the irregular shape of clay particles (particles that are often used as a cheap inert diluent) with polymers allows greater control of the release of the active ingredient, but it also improves the flow of granules through application equipment. A rotary metering system controlled by the forward speed of travel is the preferred method of optimising the dispensing of granules from a hopper. The hopper has a sloped floor to allow the granules to fall on to the metering device. The distribution from the metering device is usually through a near-vertical tube, but may be associated with an airflow from a blower.

Granules allow precise placement of a pesticide in the root zone of seedlings, and so are often attached to planters. In integrated pest management, the avoidance of spraying young seedlings has distinct advantages, as this minimises effects on natural enemies. As an alternative to soil-applied granules, root zone injection pulsed jets can create a cavity 20–25 cm deep without disturbing the surface, and have been used in turf management (Turner, 1996).

Large granules or pellets may be combined with an attractant (food, pheromone) to encourage the pest to ingest the pesticide. This technique has been used extensively in slug control. The problems are associated with maintaining the cohesiveness of the pellets even under wet conditions, so that they remain attractive to the pest, but are also not ingested by non-target organisms, particularly birds.

LURE AND KILL

Historically, sugar baits, usually as molasses, have been added to certain sprays applied as spot treatments to control the adult stage of pests, or to act as a monitoring system to detect immigration of moths. Protein hydrolysate has been used as the bait for fruitflies. Application of the bait to alternate trees along a row or even fewer trees in an orchard is usually sufficient to attract the flies and maintain control, in contrast to cover sprays which adversely affect the natural enemies. The synthesis of pheromones now allows their incorporation with an insecticide to attract the pest more selectively to a very localised application. The treatment of sticks with a mixture of insecticide and pheromone has been useful in controlling the boll weevil (*Anthonomus grandis*) by attracting the emergent adults before they reached cotton crops.

BIOLOGICALS

Despite the major changes in pesticides, the public remains concerned about their use, so research has examined the possibility of applying certain biological products as pesticides. Particular attention has been given to the use of baculoviruses, fungi and entomopathogenic nematodes. Epidemics of naturally occurring baculoviruses are usually sporadic and depend on a large host population, so efforts have been directed at mass production of the viruses so that they can be applied like a conventional chemical pesticide. Unfortunately, purified virus is adversely affected by UV light and in some cases by the pH of a leaf surface, so persistence is low and necessitates formulation with UV screens. Another problem is that the virus must be ingested, and tends to be slow acting. In consequence, the use of baculoviruses has been constrained and largely limited to forestry, for example the control of pine beauty moth *Panolis flammea*, where some pest damage can be tolerated (Entwistle, 1986).

Similar problems have beset the development of mycopesticides, as fungi require particular conditions to be effective against the host. One success was achieved with *Verticillium lecani* against aphids on chrysanthemums, as the beds of plants were enclosed in black polythene to control daylength and initiate flowering, and this raised the humidity sufficiently for the fungal spores to invade the aphids.

In developing the use of another fungus, *Metarhizium flavoviride*, to control locusts, the lack of water in desert areas led to the initial search for a formulation that would deliver spores and allow them to germinate on locusts at very low humidity (Bateman, 1997). By formulating the conidial spores in dry vegetable oil, the mycoinsecticide has been shown to be effective when applied as a ULV spray with rotary atomisers under field conditions in Africa against a number of acridid species as well as the desert locust *Schistocerca gregaria*, albeit more slowly than with certain chemicals. This success is based on delivering the appropriate concentration of spores in the optimum droplet size and formulation to ensure that sufficient spores are deposited directly on the insects or will be ingested by eating treated foliage soon after treatment.

In conventional hydraulic spraying, applying a wide droplet spectrum. much of the biological agent is wasted in the very large droplets and conversely the particulate spores may not be delivered in very small droplets, so a narrow droplet spectrum selected for optimal delivery is required and the concentration of spores adjusted to deliver sufficient spores. A more complex problem is with the application of myco-herbicides due to the number of pathogenic fungal spores needed to overwhelm the weeds natural defence mechanisms. Adequate moisture on the leaf surface ie a sufficiently long dew period is important so attention has been given to invert emulsions, but further research is needed.

Entomopathogenic Nematodes

Nematodes as a control agent have been used widely for controlling a range of insect and mollusc pests, but success has often been confined to soil treatments applied with large volumes of water. However, their use against targets other than in the soil have been tried with increasing success. Recent work has been directed at assessing new techniques for application to improve control of pests in a foliar environment.

Previously foliar treatments have given variable control even with very high volumes of water and carefully control application environments. More recent work has shown that consistent foliar application may be attainable, providing that there is an appropriate use of adjuvants to control desiccation (Glazer *et al.*, 1992), UV damage (Gaugler *et al.*, 1992; Glazer *et al.*, 1992) and to improve targeting. The use of *Steinernema feltiae* against *Liriomyza* leafminer is a good example. Initial work was undertaken as part of a wider project, and a large amount of data was collected to identify nematode behaviour on the leaf surface (Williams and Walters, 2000; Head *et al.*, 2000). From this a suitable application was developed and a product launched to enable hot-spot treatments particularly suited as part of an integrated system for complete eradication of statutory leafminer with successes of up to 99% larval mortality (Head, 2001). Later studies using flowable gels to apply the entomopathogenic nematodes to the foliage indicated the potential to reduce the volume of water required, while maintaining relatively high pest mortalities. Such gels have been assessed in small-scale trials for general nematode foliar application (Baur *et al.*, 1997), but have recently gained use more widely for other applications, such as bait stations for cockroaches (Pye *et al.*, 2001).

Polymer formulations have been suggested to enable application of IJ to foliage in a protected state, for example, nematode encapsulation in sodium alginate immediately before application has been used, so that, at the target surface, the nematode-in-gel formulation is still viscous (Navon *et al.*, 1998). However, the separate application may prove difficult technically, due to rheological problems.

These developments suggest a move away from the restrictions of high-volume soil application and their use in other situations, many of which have not been successful with traditional biological control. The use of often-massive, inundative releases utilising similar application systems to those used with chemical insecticides, is an initial move towards greater use of biopesticides, for which there are now many products on the market. In the extreme, aerial application of concentrated nematode suspensions has been successful for the control of cranberry girdler, with sufficient water supplied as part of the irrigation process via appropriate timing of nematode application.

Other novel delivery systems include the use of insects to disseminate a baculovirus, for example, to control Rhinoceros beetle, male beetles were trapped, treated and then released to spread fungal spores through a pest population.

ACKNOWLEDGEMENT

I thank Simon Piggott for assistance with the section on entomopathogenic nematodes.

REFERENCES

Bateman, R. (1997) Methods of application of microbial pesticide formulations for the control of grasshoppers and locusts. *Memoirs of the Entomological Society of Canada* **171**, 69–81.

Baur, M.E., Kaya. H.K., Gaugler, R. and Tabashnik, B. (1997) Effect of adjuvants on entomopathogenic nematodes and efficacy against *Plutella xylostella*. *Biocontrol Science and Technology* **7**, 513–525.

Clarke, B. (1988) Seed coating techniques. In: Martin T. (ed.) *Application to Seeds and Soil*. BCPC Monograph 39, pp. 205–211.

Clayton, P.B. (1993) Seed treatment. In: Matthews, G.A. and Hislop E.C. (Eds) *Application Technology for Crop Protection*, CAB International, Wallingford UK, 329–349.

Coffee, R.A. (1981) Recent developments in electrodynamic spraying. *Proceedings, 1981 British Crop Protection Conference – Pests and Diseases*, pp. 883–889.

Craig, I.P., Matthews, G.A. and Thornhill, E.W. (1993) Fluid injection metering system for closed pesticide delivery in manually operated sprayers. *Crop Protection* **12**, 549–553.

Derting, C.W. (1987) Wiper application. In: McWhorter, C.G. and Gebhardt, M.R. (Eds) *Methods of Applying Herbicides*. WSSA Monograph 4, 207–229.

Entwistle, P.F. (1986) Spray droplet deposition patterns and loading of spray droplets with NPV inclusion bodies in the control of *Panolis flammea* in pine forests. In: Samson, R.A., Vlak, J.M. and Peters D (Eds) *Fundamental and Applied Aspects of Invertebrate Pathology*, 613–615.

Gaugler, R., Bednarek, A. and Campbell, J.F. (1992) Ultraviolet inactivation of heterorhabditid and steinernematid nematodes. *Journal of Invertebrate Pathology* **59**, 155–160.

Glazer, I., Klein, M., Navon, A. and Nakache, Y. (1992) Comparison of efficacy of entomopathogenic nematodes combined with antidesiccants applied by canopy sprays against three cotton pests (Lepidoptera: Noctuidae). *Journal of Economic Entomology* **85**, 1636–1641.

Halmer, P. (1988) Technical and commercial aspects of seed pelleting and film-coating. In: Martin, T. (Ed.) *Application to Seeds and Soil*. BCPC Monograph 39, 191–204.

Head, J. (2001) Undermining the leafminer. *Grower* **40**, 20–21.

Head, J., Walters, K.F.A. and Langton, S. (2000) The compatibility of the entomopathogenic nematode, *Steinernema feltiae*, and chemical insecticides for the control of the South American leafminer, *Liriomyza huidobrensis*. *BioControl* **45**, 345–353.

Jeffs, K.A. and Tuppen, R.J. (1986) Application of pesticides to seeds. In: Jeffs, R.A. (Ed.) *Seed Treatment*, BCPC, Farnham, UK, pp. 17–45.

McAuliffe, D. (1999) Flow control of lever operated knapsack sprayers with CFValve. *International Pest Control* **41**, 21–23, 28.

Matthews, G.A. (1999) *Application of Pesticides to Crops*. IC Press, London.

Navon, A., Keren, S., Salame, L. and Glazer, I. (1998) An edible-to-eat calcium alginate gel as a carrier for entomopathogenic nematodes. *Biocontrol Science and Technology*, **8**, 429–437.

Pye, N., Holbrook, G. and Pye, A. (2001) New bait station with *Steinernema carpocapsae* nematodes kills cockroaches. *34th Annual Meeting of the Society for Invertebrate Pathology, 25–30 August, 2001*, Society for Invertebrate Pathology, Raleigh, NC, USA.

Turner, R. (1996) Practical aspects of pesticide application in amenity turf. *Pesticide Science* **47**, 386–387.

Williams, E.C. and Walters, K.F.A. (2000) Foliar application of the entomopathogenic nematodes *Steinernema feltiae* against leafminers on vegetables. *Biocontrol Science and Technology* **10**, 61–70.

6 The Aerial Application of Pesticides

N. WOODS
Director, The Centre for Pesticide Application and Safety (CPAS), The University of Queensland, Gatton Campus, Queensland 4343, Australia

INTRODUCTION

Many agricultural and public health operations are carried out around the world using aircraft. From controlling mosquitoes in Florida to eradicating mice in Australia, specialised aircraft perform tens of thousands of operations every year. In the United States, some 4100 aircraft are registered for aerial application. It is estimated that approximately 25% of the 125 million hectares harvested in the USA in 1998 were treated with crop protection products using aircraft. In Australia, some 300 aircraft are used to apply about 20% of the local crop protection chemical market to an average of about 10 million hectares annually.

In many countries, however, the use of agricultural aircraft has significantly diminished over the last 20 years. With some, the very mention of agricultural aircraft or 'crop dusting' (an incorrect term) spawns a reaction and fear of widespread pesticide drift. In Europe, the use of agricultural aircraft has declined rapidly over the last 20 years. The number of aircraft in the UK has been reduced from more than 100 aircraft in the mid-1970s to about a dozen today. So what are the facts, and how can application using airborne platforms be optimised?

The case for and against using agricultural aircraft is illustrated in Tables 6.1 and 6.2. The use of specialised agricultural aircraft developed post-war, largely as a result of the greater speed, better timing and efficiency of application offered by aerial distribution. Crossing the ground in excess of 200 km/h, aircraft are able to apply agricultural or public health products rapidly over large areas within narrow optimum application windows. When crop height and waterlogged areas restrict the passage of wheeled vehicles, aircraft are able to place pesticides strategically on crops in response to economic thresholds, without contributing to soil compaction and structural breakdown.

Because large areas of land can be covered quickly, some pesticide formulations are able to be applied uniformly at rates as low as 2 litres/ha (undiluted Ultra Low Volume (ULV) formulations) and 1 kg/ha (e.g. some sand granule formulations of insecticide).

Optimising Pesticide Use Edited by M. Wilson
 ISBN: 0-471-49075-X

Table 6.1 Some advantages of using agricultural aircraft for pesticide application

1. Aircraft can be used over wet/irrigated areas impassable to a wheeled vehicle
2. Being clear of the ground, soil compaction and wheel marks are eliminated
3. Aircraft are faster and more fuel efficient
4. Airborne application allows timely treatment of pests and diseases
5. Better coverage and penetration of a crop can be achieved in some circumstances
6. Grower labour is reduced
7. Aircraft can overcome limitations when crop height acts against ground-based equipment
8. A grower or client can confer application to skilled professional operators equipped with correctly calibrated equipment and dedicated pesticide handling systems

Table 6.2 Some disadvantages of using agricultural aircraft for pesticide application

1. Aircraft (helicopters and aeroplanes) are visible and audible and attract attention. They may cause noise pollution
2. Compared to some ground equipment, aircraft release sprays from greater heights. This may increase the drift potential
3. Low flying poses significant risks for the pilots of agricultural aircraft
4. Aircraft operations sometimes have to be confined to optimum 'application windows', which may be very short (e.g. 2–3 hours after sunrise)
5. Small cropping areas and those surrounded by obstructions or susceptible areas may not be able to be treated
6. Under/over-dosing is more likely than with ground-based application

AIRCRAFT TYPES

An agricultural aircraft is an aeroplane or helicopter used in agriculture, forestry and public health, to control pests, apply crop protection products and fertilisers, or assist in the management of livestock over large areas. After World War II, many 'surplus' aircraft, such as the Tiger Moth (in Europe), the Stearman and even twin-engined DC3 aircraft in the United States, were adapted and modified to apply liquid and solid formulations of pesticides to crops. However, demand soon led to the design and development of purpose-built aeroplanes that could lift substantially higher payloads, operate off 'rough' agricultural airstrips and incorporate a range of specific safety features, such as reinforced cockpits.

Large numbers of designs of purpose-built agricultural aircraft were developed and manufactured in the USA. In 1957, the prototype Piper Pa 25 Pawnee first took to the air. This fabric-covered aircraft proved to be safe and economical. Fitted with a 235 hp engine, the aircraft had a useful load of about 600 kg. Similarly, the Cessna C188 Ag Truck was produced in large numbers. Fitted with a 300 hp engine, this low-wing strut-braced aeroplane had a useful load of 482 kg. Larger aircraft, such as the Air Tractor AT301 and Ayres Corporation S2 Thrush, were designed in the 1970s to carry loads up to 1540 kg using large 600 hp radial engines such as the Pratt & Whitney R-1340. Such aircraft were the mainstay of many agricultural fleets in the Western world through the 1970 and 1980s. However, during the late 1980s and 1990s many operators replaced piston-engined aircraft with turbine-powered airframes. Reliable turbine power plants, such as the Pratt & Whitney 600 hp PT6A turboprop, are significantly lighter than comparable reciprocating engines. Turbine engines can be costly to purchase and overhaul. If, however, aircraft utilisation can be maintained, direct operating costs can be competitive. Higher payloads can be carried, and significant increases in range and endurance achieved, using turbine-powered aeroplanes.

HELICOPTERS

Rotary-wing aircraft are also used in agriculture. Their advantages and limitations compared to fixed wing aircraft are summarised in Tables 6.3 and 6.4.

Fundamentally, helicopters are able to manoeuvre at slower flying speeds when compared to fixed-wing aircraft, and thus are able to operate in more confined spaces. A consequence of a slower flying speed is that the airspeed over nozzle systems is reduced, and this can result in the production of larger droplet sizes, particularly from hydraulic nozzles. This can be advantageous for the application of some herbicides. However, these advantages must be balanced against the higher costs and lower payloads associated with helicopter operations.

Table 6.3 Some advantages of using helicopters for pesticide application

1. Helicopters generally operate at lower ground speeds and have greater manoeuvrability
2. Where large droplets have to be applied from hydraulic nozzle, lower flight speeds may reduce droplet shatter
3. They are better able to treat small and irregular treatment areas
4. Helicopters are versatile, able to carry personnel to the site of operations and land close to operation areas
5. Helicopters can be loaded and refuelled at the side of the treatment area
6. Helicopters have faster turning times and this may contribute to higher productivity
7. If flight forward speeds are low, the main rotor may contribute to greater crop penetration

Table 6.4 Some limitations of using helicopters for pesticide application

1. Compared to similar sized fixed wing aircraft, helicopters usually have higher purchase prices, maintenance and operating costs
2. Helicopters usually have a lower productivity in terms of the cost of operation. They tend to lift less material for a given cost and distance travelled, (less payload per engine horsepower)
3. Compared to similar sized fixed wing aircraft. Helicopters have a slower flying speed. The cost of ferrying to an operational area can be expensive

AERIAL APPLICATION TECHNOLOGY

The central aim of good application practice is that any pesticide used in agriculture or public health is applied such that the maximum deposit is applied to the target site and contamination of off-target areas is minimised. However, it is recognised that most forms of pesticide application produce small quantities of pesticide drift. In a recent notice issued by the US Environmental Protection Agency, it was concluded that the 'EPA recognizes that some *de minimus* level of drift would occur from most or all applications as a result of the uses of pesticides' (US EPA Draft Pesticide Registration Notice 2001-X).

Fortunately, in comparison to the number of missions flown by agricultural aircraft, the incidence of significant drift incidents is low. It is the responsibility of all applicators to ensure that the aerial application of pesticides is optimised to avoid any off-target movement of pesticide which could have the potential to cause adverse economic impact or damage to the environment.

Figure 6.1 gives a general overview of the main techniques adopted for the aerial application of pesticides and fertilisers.

Solid formulations can be applied straight from an aircraft or through a range of spreading devices. Fertilisers such as superphosphate, which are used for pasture improvement in some countries, are typically applied without spreading devices – the unevenness in distribution being compensated by frequent applications over several seasons. Where precise deposition of material is required, e.g. when distributing herbicides or insecticides, a carefully calibrated spreading device is usually used to ensure that material from subsequent passes of an aircraft is correctly overlapped and a uniform deposit is achieved over the ground.

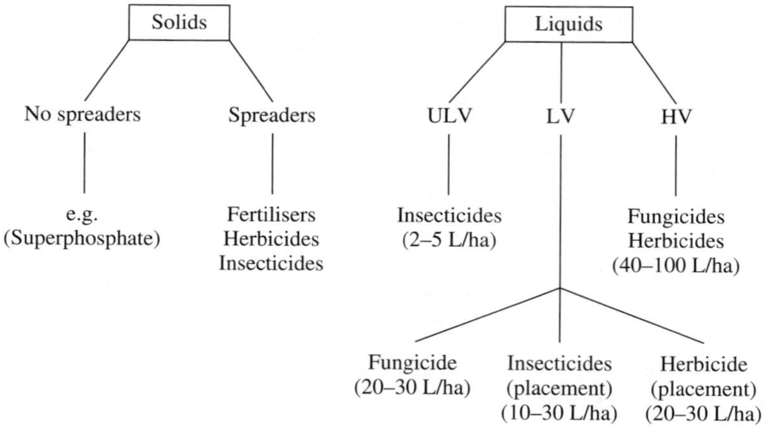

Figure 6.1 Main aerial application techniques used for applying pesticides and fertilisers. Application volumes are typical of those used in Australia

For optimisation, accurate flow-rate control, spread patterns, flying heights and flight-lane separations (or track spacings) have to be maintained.

Most pesticide products, however, are formulated as liquids. These enable small quantities of biologically active ingredients to be dispersed evenly and precisely across a target area, as long as atomisation of the liquid by a nozzle system is both adequate and uniform. With suitable nozzle systems, liquid formulations of pesticides and fertilisers are usually applied at rates varying from one to greater than 100 litres/ha, depending upon the target site, the mode of action of the product and on environmental considerations.

Three main transport processes can be used to describe the distribution of a liquid biocide from an aircraft:

1. Droplet generation (creating a large number of droplets).
2. Droplet transmission (the movement of the droplets from the nozzle through the air to the target).
3. Droplet capture (the impaction of droplets on a target).

For the application of most liquid products, the droplet size of the emitted spray is the most important criterion. As is highlighted in the following sections, coverage levels, drift profiles and efficacy are all strongly influenced by the size of the droplets generated by an aircraft. Consequently, for optimisation, a detailed knowledge of nozzle performance and the characteristics of formulations is required.

GENERATING DROPLETS FROM AIRCRAFT

Unfortunately, no practical spray nozzle produces droplets that are all the same size. All commercial spray nozzles generate a range of droplet sizes, depending upon their design formulation type and characteristics.

Nozzles that are currently in use on aircraft around the world can be classified into two main groups. Firstly, those that depend essentially on centrifugal energy to break a sheet of liquid into droplets, e.g. the Micronair, the Beecomist, and the Unimizer nozzle systems, and secondly, nozzles that use hydraulic pressure to develop a liquid sheet for droplet generation, e.g. Spraying Systems and CP Products. Some of these nozzle systems also use high-speed air and electrostatic processes to further influence atomisation and the flight behaviour of sprays, e.g. the Spectrum (ES) electrostatic nozzle.

Laser technology has enabled the rapid assessment of nozzle performance to be undertaken during the last 15 years. By simulating the airflow profile around both centrifugal and hydraulic nozzles in a wind tunnel, the influence of airspeed, formulation and nozzle design has been extensively studied.

Figure 6.2 shows the droplet spectra determined for three aircraft nozzle systems in a specialised wind tunnel at the University of Queensland, Australia. Using a Malvern laser diffraction analyser, the percentage volume in each droplet

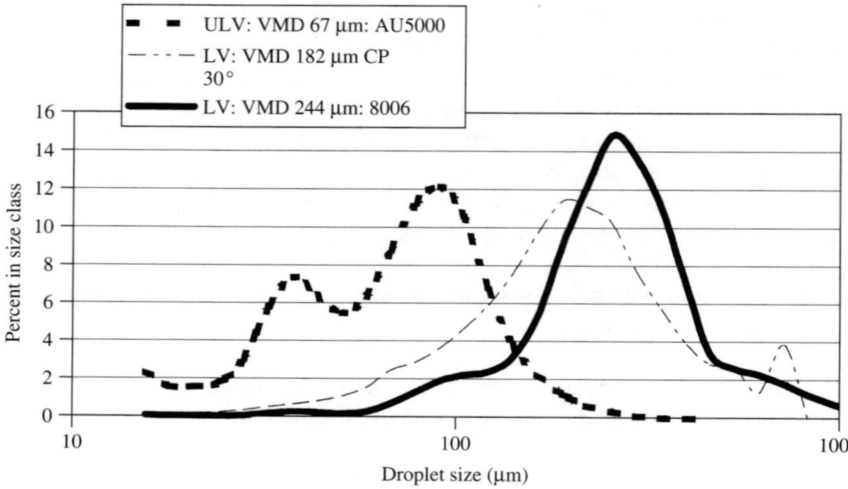

Figure 6.2 Droplet spectra (ULV and LV) generated in a wind tunnel using an AU5000 (at 51 m/s) CP and Spraying Systems 8006 flat-fan nozzle (at 67 m/s)

size class is illustrated. Both the range of droplets generated and the spread of volume median diameters (VMD), produced by a centrifugal energy Micronair AU5000, hydraulic CP nozzle and a 8006 flat-fan nozzle are illustrated.

In general terms, very small droplet sizes (<50 μm VMD) tend to be used for adult insect spraying. Small droplets (<150 μm VMD) are used for (ULV) insecticide spraying of foliage, and medium and large droplets (200–400 μm), are used for the application of insecticides, fungicides and herbicides to a range of cropping systems. Where very low-drift application is required, nozzle systems have been designed that generate droplets greater than 800 μm VMD, e.g. Microfoil and Through Value Boom (TVB) nozzles.

Table 6.5 shows droplet size data obtained in a wind tunnel with a commercial insecticide formulation, an emulsifiable concentrate (EC) and Ultra-Low Volume (ULV) formulation of endosulfan. The formulation, mixture and any added adjuvants can have a significant effect upon droplet production, and should be evaluated where specific spectra are required.

As indicated above, the droplet spectra generated by a nozzle system fitted to an aircraft is also influenced by the speed of the air passing over the nozzle tip or issuing point. This can be a significant effect and is illustrated in Figure 6.3 (Kirk 1997). These wind tunnel data show clearly that as the energy conveyed by rising air velocity increases, droplet size (VMD), is reduced and the percentage of small droplets (under 100 μm) produced by a nozzle can be increased.

Some manufacturers now routinely generate such product- and nozzle-specific data to assist aerial operators to improve the configuration of aircraft systems for specific application and target requirements. However, the relationship between

Table 6.5 Droplet spectra generated by hydraulic nozzles in a wind tunnel. Measuring device: Malvern 2600 laser diffraction analyser

Nozzle type	Size	Deflector setting	Orientation to airstream (deg)	Press (kPa)	Flow rate (L/min)	Air Speed (m/s)	Size (μm)			Span
							D[v.0.1]	D[v.0.5] VMD	D[v.0.9]	
Fan	9510		0	280	3.7	51	167	331	537	1.12
Fan	9510		45	280	3.7	51	87	202	349	1.29
Fan	9510		90	280	3.7	51	64	151	258	1.28
Fan	8006vs		0	320	2.5	67	122	243	431	1.27
Fan	8006vs		45	320	2.5	67	59	142	217	1.11
Fan	8006vs		90	320	2.5	67	43	109	173	1.19
C.JET	RF5		0	250	1.9	51	146	287	473	1.14
C.JET	RF5		45	250	1.9	51	77	185	303	1.21
C.JET	RF5		90	250	1.9	51	66	160	260	1.21
CP	0.062	coarse		145	1.8	51	135	279	474	1.21
CP	0.062	fine		145	1.8	51	67	165	274	1.25
CP	0.062	coarse		250	2.4	67	86	187	299	1.14
CP	0.062	fine		250	2.4	67	45	109	176	1.20
CP	0.078	coarse		255	3.7	51	126	275	463	1.23
CP	0.078	fine		255	3.7	51	64	157	255	1.21
CP	0.125	coarse		110	5.6	51	128	288	520	1.36
CP	0.125	fine		110	5.6	51	73	178	284	1.19
CP	0.125	coarse		85	4.9	67	76	182	358	1.55
CP	0.125	coarse		175	7.3	67	82	195	374	1.50
CP	0.125	fine		85	4.9	67	53	128	227	1.36
CP	0.125	fine		180	7.3	67	50	125	212	1.29
CP	0.172	coarse		95	7.3	67	74	188	432	1.91
CP	0.172	fine		95	7.3	67	58	140	250	1.37
Hollow cone	D6/46		0	270	3.7	51	146	304	454	1.02
Hollow cone	D6/46		45	270	3.7	51	106	236	406	1.27
Hollow cone	D6/46		90	270	3.7	51	67	164	282	1.32
Hollow cone	D6/46		0	490	4.9	67	98	214	350	1.18
Hollow cone	D6/46		45	490	4.9	67	63	155	281	1.40
Hollow cone	D6/46		90	470	4.9	67	52	126	214	1.28
WD-RD 30-25			0	90	4.9	67	94	237	580	2.05
WD-RD 30-25			45	90	4.9	67	70	172	344	1.60
WD-RD 30-25			90	90	4.9	67	60	149	280	1.47

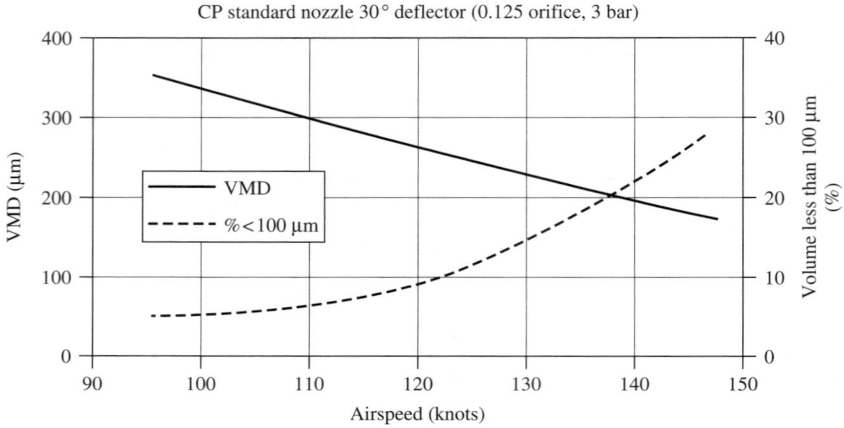

Figure 6.3 The influence of airspeed on the size of droplets generated by a hydraulic (CP) nozzle (Kirk, 1997)

the input parameters can be complex. For some nozzle types, envelope studies have been completed in wind tunnels to enable the construction of algorithms capable of describing droplet size as a function of nozzle settings and airspeed. Such equations can then be incorporated into spreadsheets and used to set up aircraft spray systems according to specific application requirements. Examples of such analysis can be found at several web sites (e.g. www.cpproductsinc.com). In such models, entering the airspeed, the angle of the air to the nozzle tip and nozzle type generates primary droplet size statistics.

APPLICATION OF LIQUIDS – FROM CLOUDS TO RAIN

The aerial application of ULV insecticides to cotton is a remarkable process. As little as 2 litres of product per 10 000 m^2 is routinely applied in countries like Australia for the control of *Helicoverpa* spp. in young and mature cotton canopies. Droplets of about 80 μm VMD are applied using AU5000 nozzles fitted to turbine Air Tractors and Thrush aircraft. Mass-balance analysis has shown that approximately 60% of the applied product is recovered by a mature canopy. With most active ingredients this is usually sufficient to achieve control. Optimisation of application during the 1980s concentrated upon increasing the amount of product applied to the target surface (upper canopy). Droplet size was decreased and volumes reduced, allowing high productivity levels to be achieved.

With fewer take-offs and landings and less ferry time being required per applied volume, ULV application has served agriculture well, particularly in areas where cropping systems are located in semi-remote regions and characterised by the use of large fields.

The fine droplets contained in a ULV spray are not all blown large distances downwind. If ULV application is undertaken such that mechanical turbulence is being generated in a neutral atmosphere, the peak pesticide deposit in the crop is located relatively close to the release point (for example 10–20 m from the line of flight of an aircraft). The droplet cloud is expanded by the turbulence and brought quickly down towards the ground. However, a downwind deposit tail is also formed and ULV technology may not be appropriate where susceptible off-target areas are located downwind.

By way of example, Figure 6.4 shows a ULV deposit pattern recovered from the top canopy of a mature cotton canopy, using fluorometry. A single application run of a ULV insecticide was applied to the cotton in a 90° cross-wind at 4 litres/ha, using a Turbine Thrush fitted with 10 Micronair AU5000 units. The droplet spectrum was measured in the laboratory prior to the experiment and found to have a VMD of 90 μm. A small quantity of UV fluorescent tracer was added to the pesticide mixture and subsequently extracted from leaf surfaces after application to illuminate the deposit pattern. Although the wind speed was 2.6 m/s (9.4 km/h), the peak insecticide deposit was located about 10 metres from the flight line. The downwind tail of droplets is clearly shown. Such patterns can be overlapped using computers to determine optimum Flight Lane Separations (FLS) and deposit data.

By way of contrast, where drift control is the overriding consideration, pesticides are applied by air, using 250–400+ μm diameter droplets (VMD) usually generated using hydraulic nozzles. Large droplets have a high kinetic energy and fall towards the ground under the influence of gravity at high terminal or sedimentation velocities. If large droplets are produced by an aircraft with the aim of laying down a uniform deposit over the ground or surface of a crop, this is termed 'placement spraying'.

Placement spraying is widely used in the United States for applying most products using water volumes of around 100 litres/ha. In Australia, herbicides

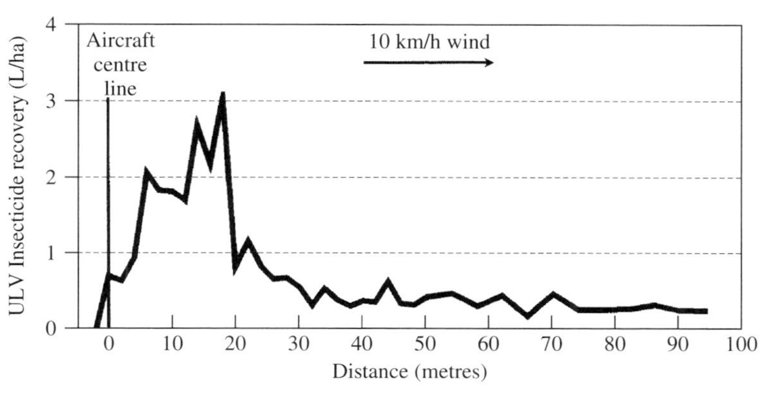

Figure 6.4 Single flight line ULV deposit pattern from the top canopy of a mature cotton crop

are routinely applied using 30 litres/ha of water and in New Zealand herbicides in forestry are applied at 50 litres/ha using large orifice nozzles angled back at 0° to the airflow. Helicopters fitted with hydraulic flat-fan (e.g. 6520) and hollow-cone (D12s no core) nozzles are used.

Technology once primarily adopted for herbicide application is now being increasingly used for low-volume (LV) insecticide application in the Australian cotton industry, in an attempt to reduce the potential for the off-target movement of sprays, and preserve the flexibility and use of productive and efficient ULV techniques.

To optimise aircraft for this type of application, large orifice nozzles are normally positioned within 65% of the wing span, balanced to give a uniform deposit over the ground, angled at 180° to the airflow and operated at low pressures (20–30 p.s.i.). Application is also normally planned to be undertaken during light wind conditions, and when temperatures are low and the relative humidity high.

At the opposite end of the spectrum, very fine droplets are used for mosquito adulticiding operations in Florida. Some Florida mosquito control districts rely on the night-time dispersal of fine droplets to knock down and suppress adult populations utilising a wide flight lane separation (300–600 m) and extensive drift (1–5 km). Particularly where mosquito-infested areas are too large to warrant effective application of larvicides, helicopters (e.g. MD 500, Bell 206) and fixed-wing aircraft (e.g. DC3, Piper Aztec, BN Islander) are utilised to apply a blanket of spray over wide areas. Fitted with a low number of flat-fan nozzles (e.g. 8001–8003) angled forward at about 135° to the airflow, droplets of Dibrom with a VMD of about 50–100 μm are applied. A heavy reliance is placed upon incident airflow to blast atomise the spray as it exits the nozzle.

To reduce the droplet size further for this type of application, new systems using high-pressure (2000+ psi) or air-assist nozzles capable of producing VMDs of 20–30 μm are beginning to be used in some parts of Florida. The efficiency of high rpm (10 000+) rotary nozzles combined with low rates of insecticide has also been demonstrated recently.

Aircraft are able to cover large areas of mosquito-infested countryside each night. To reduce nuisance population levels of *Ochlerotatus taeniorhynchus* and reduce the threat of diseases such as St Louis Encephalitis, Manatee County MCD, for example, regularly treats some 10 000–20 000 hectares per night using two MD-500 helicopters and a MBB BO105 helicopter.

APPLICATION OF SOLIDS – FROM MOSQUITO TO MOUSE CONTROL

The saltmarsh mosquito (*Oc. vigilax*) is a major vector species of the Ross River virus (Epidemic Polyarthritis) in coastal Queensland, Australia. During summer months, widespread spraying of larvae populations is carried out by local authorities in coastal mangrove environments in an attempt to lower mosquito population levels and reduce the incidence of disease transmission. Given the large areas of saltmarsh and mangrove that require treatment in South East Queensland (estimated to be 6600 ha each treatment), and the short time interval during which treatment has to occur, successful dispersion of insecticide requires the use of aircraft.

Helicopters have been traditionally favoured for use in this part of Australia given their inherent advantages in manoeuvrability and ease of operation in built-up areas. A helicopter is generally less cost effective than a fixed-wing aircraft in terms of payload to engine horsepower (Tables 6.3 and 6.4), but is able to perform productively where its performance characteristics can be used to advantage.

Sand granules have been used extensively for formulating the range of biological and chemical agents used for mosquito larviciding in coastal mangrove areas. To increase the payload of small helicopters, most products use application rates of 1–3 kg/ha. Such low delivery rates demand precision flying and the use of technology that will allow accurate track spacings and flow rates (through dispersal equipment) to be maintained.

Helicopters are equipped with either underslung buckets (fitted with a motor-driven impeller) or air-powered dispersal equipment, e.g. Isolair seeders. Both types of equipment must deliver precise flow rates to ensure that optimum uniformity across the ground and correct application rates are maintained. To maintain a high standard of commercial application pre- and post-flight tests are performed. Aircraft are calibrated to determine optimum flight lane separations (see below) and random sampling is undertaken in the field to determine actual rates and uniformity. These data can then be compared with efficacy-based measurements. Figure 6.5 shows a contour plot from such an analysis.

Low rates of material are also applied for the control of mouse plagues in Australia. In 1995, a severe mouse plague broke out in Australia at a critical time for the farming community. Mouse predictions threatened what for many farmers was the first cash-flow crop after four years of drought. In Queensland, as in New South Wales, the government initiated a broad-acre, in-crop control programme, using strychnine-coated wheat. In Queensland, approximately 0.25 million hectares of cropping land was treated using aircraft. The specially prepared wheat was applied at 1 kg/ha, equivalent to a dispersion on the ground of one grain per square metre. Standard ram air spreaders were used together with gate modifications designed to restrict flow. By using specialist 1 m^2 collectors, the dispersion pattern of the wheat was determined during pre-application trials to determine application rates and optimum flight-lane separations.

OPTIMISING AND FINE-TUNING AERIAL APPLICATION

As has been alluded to in previous sections, one of the most important aspects of aerial application is the production of a deposit pattern containing the correct number and size of droplets together with the correct dose of active ingredient.

The generation and subsequent dispersal pattern of droplets from an aircraft is influenced by droplet size, the aircraft wake, meteorology, nozzle positioning and

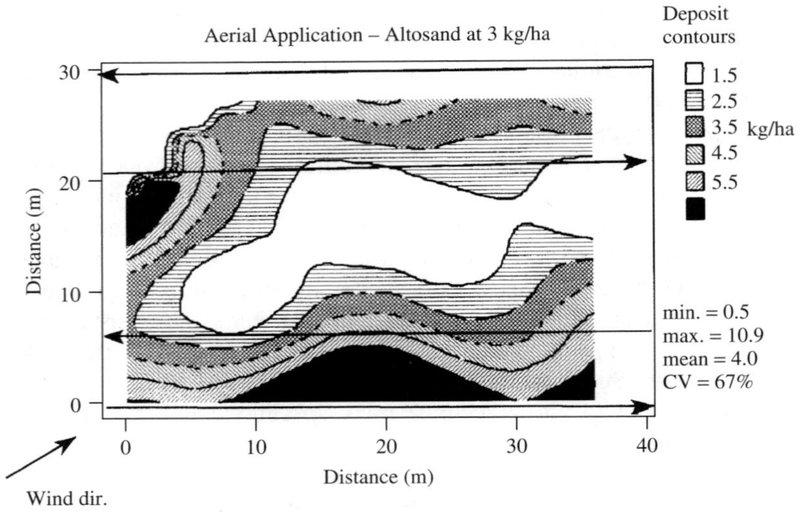

Figure 6.5 Contour plot showing the distribution of 'Altosand' granules in a coastal salt-marsh. Altosand is a commercial formulation of *S*-methoprene, an insect growth regulator used to control mosquitoes

the wing-loading/power balance. Since the flight-lane separation is a function of the distribution pattern, the selection of an optimum track spacing over the crop is dependent upon the same factors. A number of techniques are used worldwide for testing and calibrating agricultural aircraft for optimum application.

Most methods involve flying an aircraft over a target array positioned initially at right angles to the direction of flight. Testing is usually best performed under low-wind conditions (<2–3 m/s) when the effect of nozzle positioning and airflow about the aircraft is most detectable. Differences between the techniques largely centre upon target design and the methods used to analyse the deposit. (Woods, 1986).

Flat Plate Recovery

Small flat plates are positioned in rows beneath the line of flight of the aircraft. A system used in Australia comprises a series of plates mounted on a pair of stretched nylon wires. The wires are attached to a frame and can easily be moved through 360°, enabling the target array to be quickly positioned according to the direction of the prevailing wind. Such targets are best used for simulating the recovery of medium and large droplets on to soil and broadleaf plants.

Spray deposition can be monitored using water-sensitive paper or glossy white cards in conjunction with an added coloured dye. Alternatively, a fluorescent dye can be added to the tank mix and extracted from cards using solvents and quantified using a portable fluorometer.

Where a visual deposit is created, image analysis can be used accurately to quantify the deposit. Benefiting from the availability of affordable video cards, some elegant processing and measurement systems are now on the market, e.g. Droplet Technologies (see reference list for address). Suitable software linked to a flat-bed scanners has also been used.

All these systems are capable of generating deposit that can be quantified in terms of droplet number (droplets/cm^2), volume (litres/ha) or proportion of the target area covered (%). Droplet size can also be estimated (Figure 6.6).

String

A string stretched at right angles to the line of flight of the aircraft is employed by some calibration systems. In conjunction with appropriate dyes, strings can be analysed rapidly using a continuous-feed, surface-reading fluorometer. Although not quantitative, a quick comparison between treatments can be made using this technology (WRK of Arkansas).

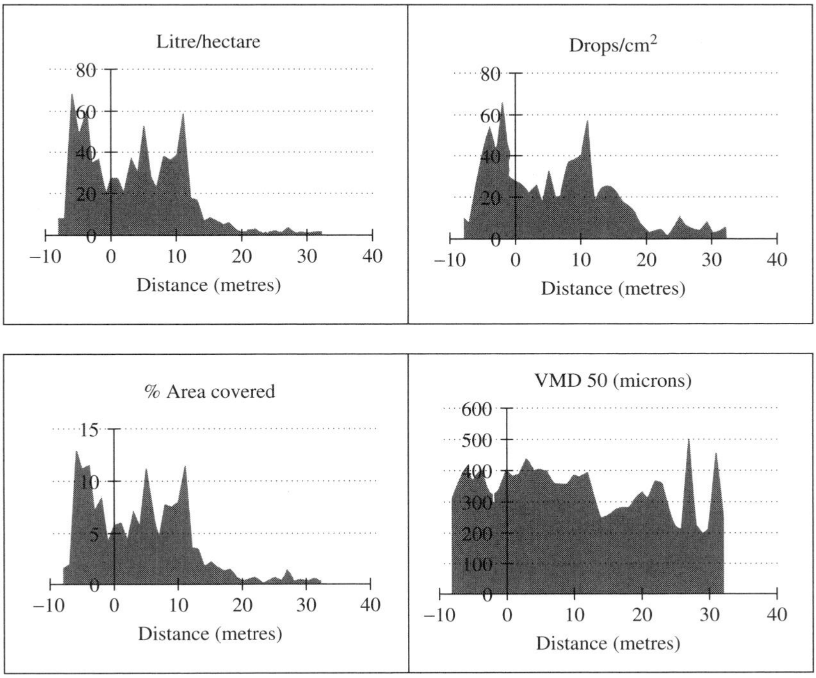

Figure 6.6 Typical droplet distribution patterns obtained from the single pass of an agricultural aircraft over a test array (Droplet Technologies)

Plant Surfaces

It must not be forgotten that a sprayed crop can be used as a spray collector. Appropriate fluorescent dyes, or the active ingredient, can be detected in crop canopies after application. The use of the crop canopy as the target is probably the best method for determining the pattern of ULV products applied to crops, as the actual target surface is used.

MAINTAINING ACCURATE TRACK SPACINGS

Once equipment settings have been determined and the correct flight-lane separation calculated, it is important that accurate track spacings are then maintained in the field. The use of satellite-based Global Positioning Systems (GPS) for providing track guidance information to pilots and accurately positioning agricultural aircraft over the paddock is now commonplace. The technology, like most computer-based systems, is developing quickly, and quietly changing the face of aerial agricultural operations around the world.

GPS now provides aerial applicators with significant capabilities. By providing aircraft tracking and guidance information, DGPS enables an operator to:

- Position an aircraft within 1–2 metres of the intended flight-lane separation.
- Apply solid and liquid material more evenly over the ground.
- Reduce the need for ground marking and therefore improve safety and reduce costs.
- Correctly identify areas to be treated while in-flight.
- Locate any gaps and unsprayed areas in a treated area.
- Keep accurate computer records of all missions, including a digital map of all treated areas.

As a simple example of the benefits of DGPS systems, Figures 6.7a and 6.7b show the actual and intended flight path of a helicopter plotted over coastal saltmarsh during a validation experiment in Australia. In the first trial (Figure 6.7a) the pilot was asked to fly the mission at a track spacing of 18 m without any reference to the GPS system (the DGPS display in the cockpit was disabled). The mean track spacing was 29.1 m. In the second test (Figure 6.7b) the DGPS was activated. Although the pilot had no previous experience of DGPS, the adherence to a nominated track spacing of 18 m (mean 18.1 m) when the DGPS system was used is clearly illustrated (Woods and Dorr, 1996).

THE ULTIMATE OPTIMISATION – SPRAY-DRIFT MINIMISATION AND CONTROL

As previously discussed, spray-drift minimisation and control is probably the biggest issue facing the users of agricultural aviation in the western world. In the

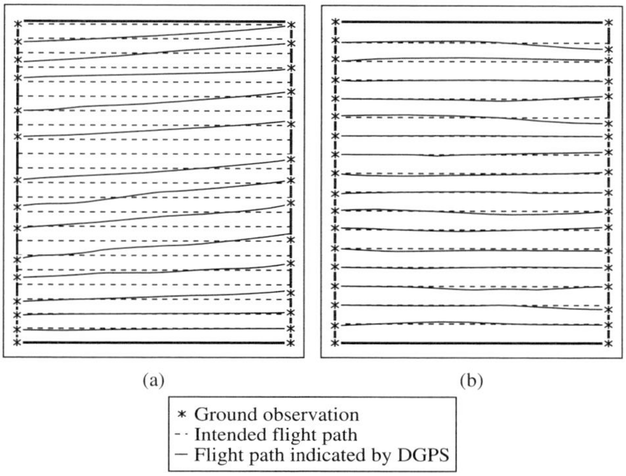

Figure 6.7 DGPS output showing actual and intended flight path of a helicopter over a coastal saltmarsh. (a) DGPS display in cockpit disabled. (b) DGPS display enabled

United States, a major research programme was initiated in 1990 by a consortium of about 39 chemical companies, the Spray Drift Task Force (SDTF). The SDTF which was formed to support spray-drift registration requirements for the US Environment Protection Agency (EPA), conducted an extensive trial programme consisting of some 300 experimental applications.

In a recent summary report, the SDTF state that from their studies droplet size was the most important factor affecting spray drift. Their studies support the view that 'the physical properties of the spray mixture generally have a small effect relative to the combined effects of equipment parameters, application technique and the weather'.

The SDTF concluded that drift levels can be minimised by (among a number of ways), applying the coarsest droplet-size spectrum that provides sufficient coverage and pest control, keeping release height low and applying pesticides when the wind speed is low (Johnson, 1997).

In Australia, a study has been undertaken in the cotton industry over the last few years to examine the aerial transport process, with the aim of minimising the impact of pesticides in the riverine environment. Significant differences in drift levels between large (*c*.250 μm) and small (*c*.60 μm) droplets were measured (Woods *et al*., 2001). The aim of the industry-focused programme has been to adopt results into a broad best-management programme that encourages drift reduction technology, while at the same time allowing growers to use ULV technology where its impact on the environment is minimal.

To analyse the complex interaction between variables, the SDTF rigorously compared downwind deposit data against predictions from sophisticated simulation models developed by the USDA Forest Service. Modifying code utilising Lagrangian equations to describe the movement of droplets in the airflow about aircraft, a new software package (AgDrift™) was developed, aimed at predicting pesticide drift levels, given a set of input parameters. It is hoped that, when fully adopted by the EPA, the model will allow evaluation of a much wider range of applications than those tested in the original field trials, and will assist in the registration of new agrochemicals.

The theoretical deposition of a ULV spray across a field is shown in Figure 6.8. In this example, a Gaussian diffusion model has been used to simulate the pesticide load arriving above the canopy as well as to predict a downwind (off-target) deposition decay curve. It is clear that if validated within described boundaries, such a modelling approach has the potential to better position an aircraft (over a crop) by estimating appropriate in-crop offset buffer zones. That is, locate an aircraft upwind of the downwind edge of a field such that most of the deposition decay curve is maintained within the target area.

PRECISION APPLICATION, THE FUTURE OF AGRICULTURAL AVIATION

So what is the future of agricultural aviation? The author remains confident that aircraft provide a viable and efficient means of pesticide delivery in a wide range of environments and situations. How then should aerial application be managed to reduce the potential threat of off-target damage and spray drift?

Although droplet size control remains the most important parameter influencing spray drift, successful management of pesticides in the field requires

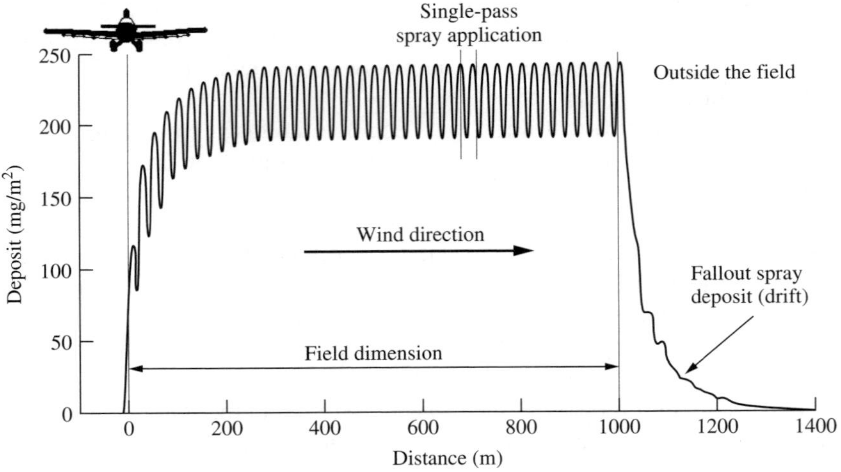

Figure 6.8 Theoretical deposition of an ULV spray across a field simulated using a Gaussian diffusion model

more from growers and applicators than just the selection of appropriate nozzles. For example:

- Communication: dialogue between pilots, growers, consultants and the wider community is essential for optimum drift control. The pre-application identification of susceptible areas, pre-season selection of appropriate wind directions and the selection of optimum application technology are arguably the most important aspects of a drift control management programme.
- Meteorology: the use of appropriate wind vectors to steer drift away from susceptible areas, awareness of the dangers of strongly stable and unstable atmospheric conditions and the selection of low-wind velocities for placement application.
- The implementation of appropriate in-crop offset buffer distances when the downwind edges of fields lie close to susceptible areas.
- Real Time Decision Making: There has been a rapid development in DGPS systems capable of receiving differential signals from space over wide geographical areas. This, together with the rise in confidence that has occurred with the development of both Gaussian and Lagrangian approaches to spray-drift modelling, is leading to the feasibility of computer assisted spray-drift management decisions being made in the cockpit.

This technology will require real-time measurement of meteorological parameters from on-board sensors or radio links to ground-based portable meteorological stations. However, it may soon be possible for agricultural pilots and the community to have confidence that an aircraft is being placed over a crop or in the urban/rural interface with full reference to environmental parameters and susceptible off-target areas.

Integrating GIS with DGPS and suitable spray-drift models will enable a computer to position an aircraft with reference to the prevailing wind direction and nozzle settings, (droplet size). Allowing for buffer zones and premapped off-target areas, real-time positioning of the aircraft will be possible. Such systems will be able to alert pilots to the likelihood and magnitude of off-target drift and even automatically prevent the inadvertent release of pesticide from an aircraft when incorrectly positioned over a crop.

It therefore may be possible that aircraft in the not too distant future will be seen to be not only the fastest and most timely way of delivering a pesticide to a target, but also the most accurate and sensitive to the environment.

REFERENCES

Droplet Technologies Inc. 937-1 West Whitehall Road, State College, PA 16801, USA.

Johnson, D., Spray Drift Task Force (1997) A Summary of aerial Application Studies Stewart Agricultural Research Services, PO Box 509, Macon, Missouri 63552, USA.

Kirk, I.W. (1997) Application Parameters for CP Nozzles, ASAE Paper AA97-006, ASAE, 2950 Niles Rd, St Joseph, MI 49085-9659, USA.

USEPA (2001) Draft Registration Notice Spray and Dust Drift Label Statements for Pesticide Products. Pesticide Registration (PR) Notice 2001-X.

Woods N., Craig I.P. and Dorr, G. (2001) Measurement of spray drift of pesticides arising from aerial application in cotton. *Journal of Environmental Quality* **30(3)**, May–June 2001, 697–701.

Woods, N. and Dorr, G. (1996) DGPS for the aerial application of pesticides *Agricultural Engineering Australia* **25(2)**, 31–33.

Woods, N. (1986) Agricultural aircraft spray performance: calibration for commercial operations. *Crop Protection*, 1986, **5(6)**: 417–421.

7 Formulating Pesticides

RUPERT SOHM
Syngenta Crop Protection, Münchwilen, CH-4333, Switzerland

INTRODUCTION

Formulation chemists in the agrochemicals industry face the challenge of delivering the full potential of pesticide active ingredients, while minimising the risk of harm to nontarget organisms. The latter include the manufacturer, the user, the consumer and the environment. Appropriate formulation design has been instrumental in bringing about significant improvements in the efficiency of pesticide use. Yet there is still considerable scope for improvements, if the technology challenges can be met.

For the sake of simplicity and brevity, this review will be focused on the application of pesticides to crops via the use of spray application equipment. The application of formulation technology to improve the efficiency of other pesticide application regimes, e.g. seed dressings, public health products, etc., will not be addressed, due to the space constraints.

Another deliberate exclusion in this review is a discussion on the application of formulation science to facilitate the safe use of the pesticide product.

THE CHALLENGES FOR FORMULATION CHEMISTS

The potential for formulation technology to improve the efficiency of pesticide applications is considerable. It begins with the application of the product, and ends with the full expression of the biological effect. Perhaps the simplest way to illustrate this is to consider the large number of obstacles that a pesticide must traverse in order to reach the biological site of action (Figure 7.1).

As can be seen from Figure 7.1, the means by which the pesticide may be diverted from its target following its application are quite varied. The following list is not exhaustive:

- Spray drift – this can result in losses of pesticide from the target site, but more concerning is the potential of the pesticide to cause harm to the neighbouring crop or environment.

Optimising Pesticide Use Edited by M. Wilson
 ISBN: 0-471-49075-X

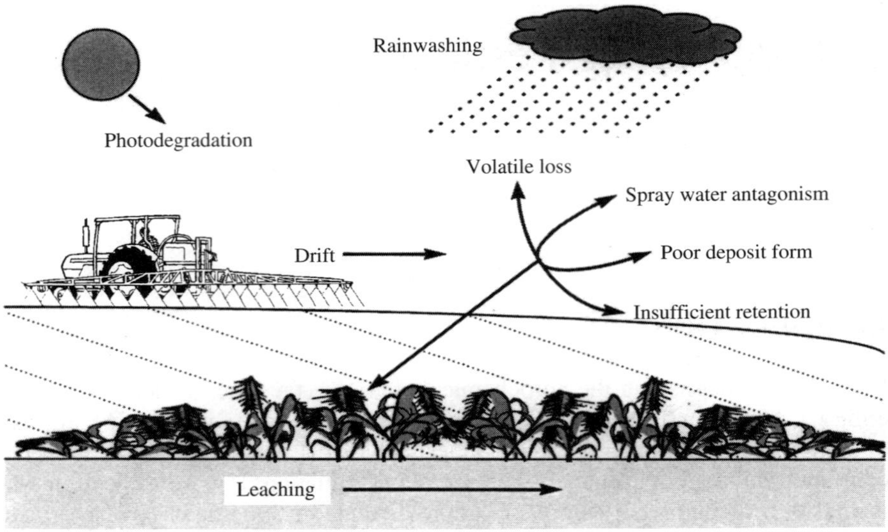

Figure 7.1 Environmental factors leading to pesticide losses

- Failure of the pesticide to be captured by the target foliage – on occasion, reduced retention is required for soil-active pesticides. In such cases, reduced foliage retention would lead to more pesticide entering the soil environment. This might be important in minimum-till or zero-till cultures, where dead foliage might prevent pesticides reaching the soil surface.
- The wide variety of means by which the pesticide might be lost from the spray deposit – degradation by sunlight, rainwashing, volatilisation, leaching, etc., are all consequences of the environment on the pesticide deposit. Quite clearly, each pesticide and use environment will bring a unique balance of priorities for the formulation chemist.
- Finally, for foliar-applied pesticides, the target plant system may impede the uptake and movement of the pesticide to its biological target.

Quite clearly, the sequential nature of the above factors means that, generally, little of the applied pesticide reaches its target. For example, only about 3.5% of the herbicide fluazifop-*P*-butyl captured by leaf surfaces actually reaches the plant meristems (Burnett, pers. comm.). This is probably an example of good efficiency in that fluazifop-*P*-butyl is not prone to a wide variety of pesticide loss mechanisms described in this chapter. There is clearly significant potential for improving the efficiency of pesticide application. Formulators of pesticides can contribute significantly to this process. As should be apparent from the above listing, the challenge is usually met by the interactions of formulators with experts from many other disciplines.

In the following sections, this wide range of potential loss mechanisms is addressed and, for simplicity, the route taken by the pesticide to the target site of action is split into three stages:

1. spraying;
2. the spray deposit;
3. movement to the site of action.

SPRAYING

SPRAY DRIFT

Possibly the first significant means by which a pesticide might be diverted from its target can occur immediately after spraying. This is the potential of the spray droplets to be diverted away from the crops or weeds by the action of climatic conditions. This process, commonly known as 'spray drift', is now becoming a greater concern, due to a number of trends in the pesticide user environment:

- The continuing use of aerial application technologies.
- The reduction of water volumes and the often consequential reduction in droplet sizes required to ensure good coverage of the target.
- The increasing use of nonselective herbicide products on herbicide-tolerant crops accompanied by the concerns on the safety of adjacent nontolerant crops or noncropped environment.

Although effective design and utilisation of the application equipment used plays by far the most significant role in minimising spray drift, intelligent formulation design can play a supporting role. The pesticide formulator's efforts are generally focused on reducing the proportion of fine droplets formed; clearly, such fine droplets tend to drift more than coarser droplets. The usual approach taken is to modify the physical properties of the spray dispersion to hinder the formation of very fine droplets at the spray nozzle.

The key properties of the spray dispersion that influence the size of the droplets formed are its dynamic surface tension and its viscosity (Elliott and Wilson, 1983; Dexter and Huddleston, 1998) although it should be noted that poor spray machinery setup and use can play a more important role than the physical properties of the spray dispersion (Spray Drift Taskforce, 1977).

The traditional formulation approach to reducing spray drift has been to increase the viscosity. The viscosity is usually increased by the use of polymers. Although there have been some advances via this route, there is still much room for improvement (Chapple *et al.*, 1993). In particular, it has proved to be very difficult to combine all the desirable properties of the additive:

- The additive should bring about a significant increase in the viscosity of the spray dispersion, yet, when formulated with the pesticide, it should not cause the pesticide to become too viscous for pouring and dispersion. In other words, how does one prevent a viscosity-increasing substance from increasing the viscosity of the more concentrated pesticide formulation? In practice, this problem is circumvented by the use of separate spray-drift retardant formulations.

- The additive should not degrade under the high shear conditions in the spray tank. More recent developments in spray-drift additives have begun to address this challenge (Hazen, 1996).
- The additive should be biologically inert and not interact adversely with the other formulation constituents (or disrupt the functions that they perform – e.g. the physical destabilisation of the pesticide formulation).

A word of caution is appropriate at this stage – the spray droplet size cannot be increased indefinitely, as ultimately overly coarse droplets will reduce spray droplet retention (Schaefer, and Allsop, 1983).

Less attention has been paid to reducing spray drift by modification of the dynamic surface tension of the spray dispersion. This is not surprising, in that a reduction in the dynamic surface tension may adversely affect the retention of the spray droplets on the leaf surfaces. Furthermore, the extensive use of tank mixes may bring about an unintended reduction of the dynamic surface tension via the contribution of the surface-active agents in the partner products.

Nevertheless, a reduction in spray retention brought about by an increase in the dynamic surface tension may be offset by increasing the elasticity of the spray droplet (Hart, and Young, 1987; Wirth *et al.*, 1991). Therefore, an approach that involves the use of polymers to both increase the viscosity and avoid the use of surface-active agents is conceivable.

Finally, spray drift caused by other factors, e.g. temperature inversions and droplet size reduction by water evaporation, cannot be addressed by the above measures (Elliott and Wilson, 1983). It is not clear whether the consequences of temperature inversion can be addressed by formulation design. However, the loss of water from spray droplets might be usefully minimised by the use of humectants.

SPRAY CAPTURE AND RETENTION

The capture and retention of the spray droplets is often reduced to addressing the retention of spray droplets on leaf surfaces. This means that the direct loss of spray droplets to the soil surface is forgotten. Figure 7.2 shows how the density of the target foliage plays an important role in determining the proportion of the pesticide captured by the target plants.

Ideally, all the pesticide spray directed at the target flora should be captured. A significant improvement in the proportion of pesticides reaching the targets was brought about in the early 1980s by the development of electrostatic spraying technology (Coffee, 1981). This approach involves the generation of charged non-aqueous droplets via an electrostatic sprayer. These charged droplets are attracted

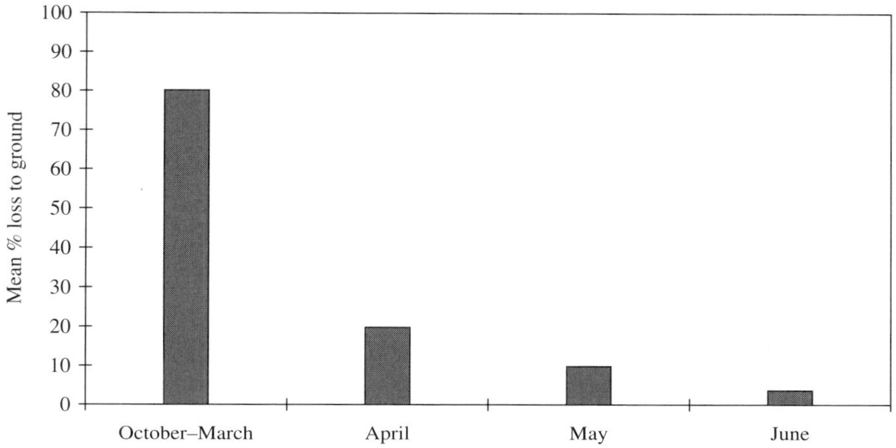

Figure 7.2 Loss of plant-protection products to soil in a winter wheat crop, as a function of growth stage (Ganzelmeier, 1999)

to the target foliage, resulting in preferential movement of spray droplets towards the target foliage and away from plant-free areas. As a result, less pesticide is lost to the soil surface. However, the inertia associated with the nearly universal use of water-based application equipment and various technical issues prevented this technology from being widely adopted. Interest in this approach might be resurrected if a lower-cost adaptation of this approach could be introduced.

In a related area, the use of oil based application media (in place of the conventionally used water) have shown improvements in the end effects achieved (Bohannen and Jordan, 1995; Hanks, 1995). Whether these are generated by the improved retention of oil droplets on the leaf surfaces, or by an enhancement of the uptake, will depend on the properties of the pesticide being studied. Once again, the inertia associated with the nearly universal use of water-based application equipment prevents this technology escaping niche uses (such as in banana plantations).

Having reached the target foliage, the spray droplets need to be captured effectively, and losses due to 'runoff' or 'bounce' need to be minimised. Addressing the capture and retention of spray droplets on foliar crops is usually a significant feature in the design of pesticide formulations. However, the formulator needs to take care when addressing the spray retention properties of a pesticide formulation. The incorporation of excessively high-levels of retention aids may be caused by over-reliance on glasshouse bioefficacy data. The glasshouse testing environment, with its coddled test plants, often provides an unnecessarily harsh view of the difficulty of wetting field-grown foliage (Anderson, *et al.*, 1987).

That the dynamic surface tensions of spray droplets are clearly an important factor in determining the degree of foliar retention has been known for a long time (Anderson and Hall, 1989). However, total reliance on this parameter has often not been appropriate, as the retention of spray droplets does not show a linear response to dynamic surface tension. Some recent work has shown that retention correlates better with a derivative of the dynamic surface tension. This derivative takes into account the rate of surfactant adsorption at the air:liquid interface and the dilational modulus of the droplet surface (Reekmans *et al.*, 1998). This development should allow the more rapid reliable screening of novel formulations for spray droplet retention properties, via the determination of simple physical parameters. On the whole, the area of spray retention has become an area where few significant improvements are anticipated. Thus, in the current state of the art, little concern is placed on understanding retention issues. The consequence of this is that formulations can now be designed to contain an optimum level of retention aid, allowing sufficient formulation 'space' for other formulation aids.

Clearly, the above considerations apply to the application of pesticides to growing crops or weeds. Soil-applied pesticides are not considered to suffer from spray droplet capture and retention issues. Indeed, the presence of foliage (dead or alive) can hinder the movement of pesticides to the soil surface. This is particularly the case in no-till applications, where the presence of stubble, etc. from the previous season's crop may act as a trap for the pesticide. In such an environment, one might attempt to develop 'anti-wetting' systems to prevent capture of the pesticide spray droplets by extraneous foliage. Although straightforward in concept, the challenge here would be to 'override' the natural surface tension reduction effects arising from the formulants present in the pesticide formulation or any tank-mix partner products. It is not clear how such a challenge might be met – especially as tank mixes of pesticides might lead to the introduction of spray retention aids. At the moment, a number of current soil-applied pesticides require soil incorporation and, in these cases, an anti-wetting system might be helpful.

THE SPRAY DEPOSIT

While the subject areas of spray drift and retention have received much attention in the past, this does not appear to be the case for all the subsequent pesticide loss mechanisms. This is partly a result of the fact that, whereas the former are relatively universal effects, not all pesticides suffer from all post-application loss mechanisms. The following list of potential loss mechanisms is not by any means exhaustive:

- Is the deposit of the appropriate size and shape?
- Will the deposit be lost due to the action of rain?
- Will the pesticide be lost due to volatilisation?
- Will the pesticide be degraded by sunlight?

- Will the pesticide be bound by ionic species present in spray solution?
- Will the pesticide leach away from the application site?

Each of these issues will be discussed in turn, although not all are relevant for both soil- and foliar-applied pesticides. Furthermore, the extent to which pesticides are vulnerable to the above loss mechanisms will depend on the precise chemistry and mode of action of the pesticide concerned.

In many cases, the end benefit of addressing the above challenges is not just a reduction in the amount of the pesticide applied, but also an extension in the period of the effect generated. Quite clearly, such changes to the profile of the pesticide are appreciated by the crop grower.

OPTIMISING DEPOSIT FORM

The challenge in optimising the deposit form lies as much in defining what the deposit form needs to be rather as in how to achieve it. One interesting example is a recent study showing how the deposit size may be tuned to provide optimal control for different insect species (Ebert *et al.*, 1999). It is not surprising that the effectiveness of contact or ingested insecticide deposits on a leaf surface will depend on the size and frequency of the deposits as well as the feeding habits of the target pest. Similarly, the uptake of glyphosate appears to be dependent on the deposit area (Knoche and Bukovac, 1993). The glyphosate deposit area debate centres around whether the smaller deposit area leads to a greater osmotic pressure gradient into the weed and/or whether the smaller deposit leads to the spray droplet drying more slowly, and thus keeping the water-soluble active ingredient in solution for a longer period of time. In either case, the formulator ignores the nature of the deposit formed on drying, at their own peril. Fortunately, the understanding that the extent of spray droplet spreading (and hence the deposit area) can be manipulated independently of the droplet's retention has been known for some time (Anderson, *et al.* 1989). This and the ongoing improvements in microscopic surface analysis should make deposit modification a fruitful area for future improvements in pesticide efficacy.

AVOIDING LOSSES DUE TO THE ACTION OF RAIN

The loss of pesticide through the action of rain appears to provide as much work for researchers as it causes the agricultural industry in its management. The degree to which this provides a cause for concern is dependent on the physical properties of the active ingredient and the time required for uptake process to remove the pesticide. Quite clearly, the climatic environment plays a significant role in determining the importance of this dimension, for example, plantation environments in South-East Asia present a significant challenge to the pesticide applicator and formulator.

Avoiding losses of glyphosate, a water-soluble active ingredient requiring many hours of rain-free uptake, is necessarily a subject of great interest. Conversely, the rapid uptake of paraquat into plant tissue means that loss of this active ingredient by rainwashing is not a cause for concern. Addressing glyphosate losses due to rainwashing has been largely driven by attempts to improve the speed of uptake through the choice of appropriate adjuvant systems. An extreme example has been the use of silicone wetters, where a stomatal infiltration mechanism has been suggested (Field and Bishop, 1988; Stevens *et al.*, 1991). However, this apparently radically different uptake mechanism for glyphosate appears to cause adverse changes in the spectrum of the weeds controlled by this herbicide (Baylis and Hart, 1993; Gaskin and Stevens, 1993). Quite clearly, there are still opportunities for improving the rain-fastness of this water-soluble active ingredient.

Insoluble particulate pesticides present a less-daunting challenge. In addition to increasing the rate of uptake, or decreasing the particle size (for better adhesion) (Maas, 1979), the use of formulation components that physically bind the pesticide particles has been extensively explored. Such additives are often available as tank-mix additives, e.g. 'Spraymate Bond', 'Nufilm P' and oil adjuvants (Kudsk, and Mathiassen, 1991; Kudsk, 1992). These products are invariably polymeric, and the challenges faced in their use involve building the product into a pesticide formulation (for a single shot product in markets where tank-mixes are not customarily used). An extreme consequence of an effective rain-fastness aid might be the irreversible capture of the pesticide, leading to reduced uptake into the target plant (or reduced exposure to the pest). However, such circumstances must be extremely rare.

REDUCING THE VOLATILISATION OF THE PESTICIDE

The loss of pesticides due to volatilisation should not be underestimated. It has been estimated that much more pesticide is lost from its target by volatilisation than from spray drift (Anon., 1996). The pesticide formulator tends to encounter this problem when working with some soil-applied pesticides, or where a leaf deposit needs to remain on the plant surface. Pesticides taken up by foliage do not tend to suffer from this problem, as plant uptake mechanisms usually ensure that pesticide soon ceases to be available to this loss mechanism.

Microencapsulation of the active ingredient is often the only option open to the formulator to address this loss mechanism. In the case of the thiocarbamate herbicides EPTC and vernolate, the application of microencapsulation technology has lead to significant reductions in active ingredient volatilisation (Groenwald *et al.*, 1980). Similarly, clomazone, an active ingredient with well-known problems associated with its volatility, has been successfully microencapsulated in order to address these problems (Lee and Nicholson, 1996). In these situations, the benefits accorded by microencapsulation technology can be translated into benefits such as the ability to avoid or delay any soil incorporation, as depicted in Figure 7.3.

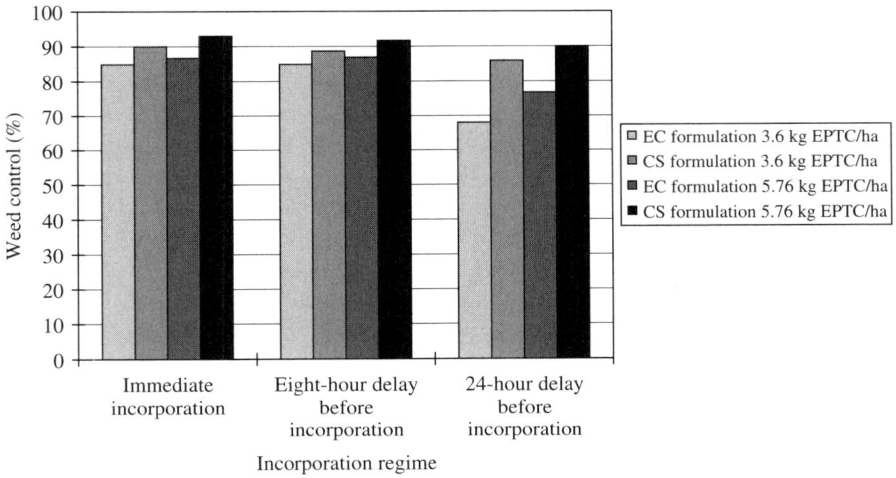

Figure 7.3 The influence of formulation type and timing of EPTC incorporation on weed control. Source: Scher *et al.* (1998)

MINIMISING LOSSES ARISING FROM SUNLIGHT-DRIVEN CHEMICAL DEGRADATION PROCESSES

The need for many pesticides to interact chemically with target sites in the pest often results in such molecules being prone to degradation through the action of sunlight. This problem is exacerbated by the need to divide the pesticide dose into very small particulates in order to generate the coverage required and to optimise the pesticide interaction with the crop, weed or pest. Such formulation design processes lead to significant surface areas per gram of active ingredient (typically of the order of $1-2\,m^2$ per gram of active ingredient). Minimising the action of sunlight on such finely divided active ingredient presents quite a challenge to the formulator. Solutions to the problem have been quite varied. Most commonly, an UV adsorbing substance or antioxidant (or a combination) is used to protect the pesticide. Aside from commercial UV stabilisers used in the plastics industry, substances such as lignins and derivatives of lignins have been used to some effect (Humphrey, 1998; Westvaco Technical Brochure, 2000) Often, care in the selection of the formulation type employed can help to optimise UV stability – see Figure 7.4.

The need to match the UV absorption spectrum of the pesticide, plus developing a solution that fits the use profile of the product (e.g. soil-applied, vs. foliar-applied, etc.) means that most solutions to UV stability are particular to the pesticide protected. A significant obstacle to the extensive use of UV stabilisers is their often significant cost (relative to the pesticide), as well as the high application rates applied. Furthermore, the improvements seen to date have

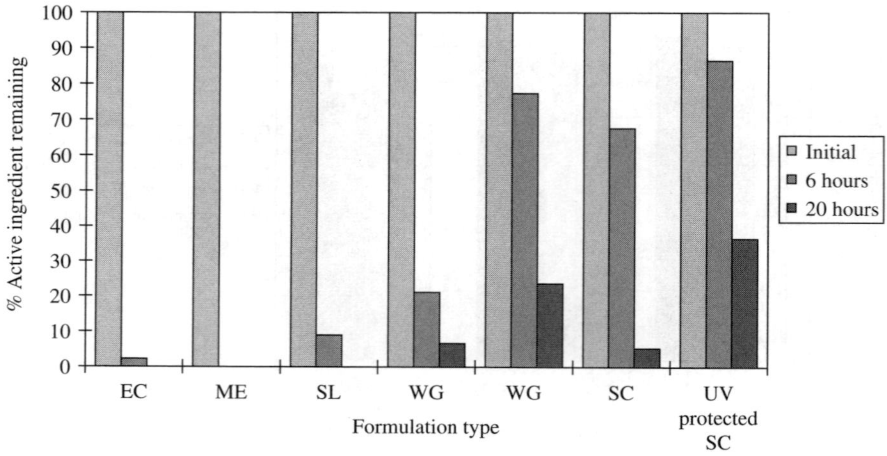

Figure 7.4 Influence of formulation type on UV – laboratory studies on a novel pesticide (Paterson, 1998)

tended to enhance the performance of many marginal pesticides – there is still a need to generate practical systems which increase the half-lives of pesticides by several orders of magnitude.

AVOIDING LOSSES IN EFFICACY DUE TO WATER QUALITY

The nominally inert medium of the spray dispersion – water – is known to antagonise the efficacy of a number of pesticides. This antagonism of pesticides arises from the calcium, magnesium – and sometimes sodium – ions found in normal waters. Such antagonism is particularly prevalent where the active ingredient itself is an acid or a salt of an acid. The presence of certain cations can lead to the formation of less biologically active salts. Although a known phenomenon with a variety of herbicides, the best-known example is the interaction of glyphosate with calcium ions to form a calcium glyphosate complex (Nalewaja and Matysiak, 1993; Nalewaja *et al.*, 1994; Thelen *et al.*, 1995). This antagonism is often resolved by the incorporation of ammonium sulphate into the formulated product, or its addition as a tank mix additive. It has been proposed that the sulphate anions effectively remove the calcium from the drying spray droplet (Nalewaja and Matysiak, 1991). Whatever the mechanism, the pesticide formulator is faced with the challenge of incorporating this, or another 'water quality modifier', into the end product at levels appropriate for all the anticipated use environments. Perhaps the most powerful aid for addressing water-quality problems is a reduction in the volumes of water applied (and hence the quantity of 'bad ions' available to cause antagonism).

MINIMISING LEACHING FROM THE SITE OF APPLICATION AND EXTENDING THE PERIOD OF CONTROL

The movement of soil-applied pesticides away from their site of action is currently a significant cause of concern (US EPA web site). In essence, there needs to be a balancing act between ensuring that the soil environment contains a sufficiently high level of active ingredient to control the target organisms, without providing a surfeit that can add no crop protection value because it only leaches away. From a commercial viewpoint, this problem is analogous to the problems encountered in extending the length of time that a pesticide application provides control of the target organisms. Conceptually, a metering system is required that applies the pesticide in a fashion that tops up the levels of free active ingredient as it decays. Generally, the only practical solution has been to formulate the pesticide as a microcapsule formulation. In the case of acetochlor, microencapsulation has extended the period of control (Scher *et al.*, 1998). Similarly, the leaching of dicamba has also been controlled by the use of microencapsulation technology (Curtis *et al.*, 1995). In the latter case, the encapsulated formulation showed a delayed effect compared with the conventional aqueous solution formulation of dicamba.

Currently, the release process from microcapsule formulations is managed by simple diffusion of the active ingredient through the polymer walls. Generally, an increase in the pesticide breakdown brought about by rising soil temperatures is partially compensated by an increase in the rate of diffusion of the pesticide through the microcapsule walls.

Nevertheless, this is a challenge which often stretches the capability of formulation scientists to generate solutions. Opportunities clearly exist for truly triggered release formulations that deliver free pesticide on the basis of a temperature, a biological or another environmental trigger.

MOVEMENT TO THE SITE OF ACTION

Having avoided the variety of loss mechanisms described above, the next stage is for the active ingredient to move to the site of action within the biological mechanism being addressed. This may involve penetration into an insect pest, a fungal growth or movement into a plant (whether a weed or a crop). This area of science aims to improve the foliar uptake of pesticides and their onward translocation. Quite clearly, the rate and extent of uptake play a significant role in determining the biological end effect. Intelligent formulation design and/or the appropriate use of 'adjuvants' can enhance the biological effect, above and beyond addressing the potential loss mechanisms described above. Significant efforts are applied to improving the efficiency of pesticide use through formulation design and, most importantly, the judicious choice of additives. In the interests of brevity, only three of many possible themes will be described, namely:

1. The impact of pesticide particle size on the efficacy of foliar pesticides.

2. The use of an 'accelerator' adjuvant to enhance the activity of a pesticide.
3. The minimisation of crop damage by the judicious choice of formulation type.

The existence of many active ingredients as crystalline solids presents a challenge to the formulator, in that this form is not appropriate for optimal uptake – the active ingredient needs to be solubilised in some way to penetrate the leaf surface. This and the need to distribute the active ingredient across large areas of soil/crop leads to the need to grind crystalline active ingredients. It has long been observed that the efficacy of many crystalline foliar active ingredients is dependent on the particle size of the active ingredient (Maas, 1979; Backman *et al.*, 1976). A typical example is displayed in Figure 7.5.

Adjuvants that enhance the efficacy of pesticides are often labelled as 'activators' (Stock *et al.*, 1992). Their action in enhancing the uptake of pesticides may be either through a passive mechanism involving no direct action on the host, or via a more active mechanism, whereby the mobility of the pesticide through the cuticular membrane of the leaf is enhanced. Ensuring that a spray deposit remains moist through the use of a humectant (e.g. glycerol) is an example of a 'passive' role that might be played by an adjuvant (Matsumoto *et al.*, 1992). Adjuvants that play an active role are often labelled 'accelerators,' and are defined as acting on the host plant so as to enhance penetration of the active ingredient (Schönherr *et al.*, 1999). One of many examples of this effect is demonstrated in the case of the product 'Flexstar' – a formulation of the herbicide fomesafen, containing a built-in 'accelerator' adjuvant. Table 7.1 shows how the presence of a tailored built-in adjuvant system can demonstrate improvements over an adjuvant-free formulation when both are used with conventional tank mixed adjuvants.

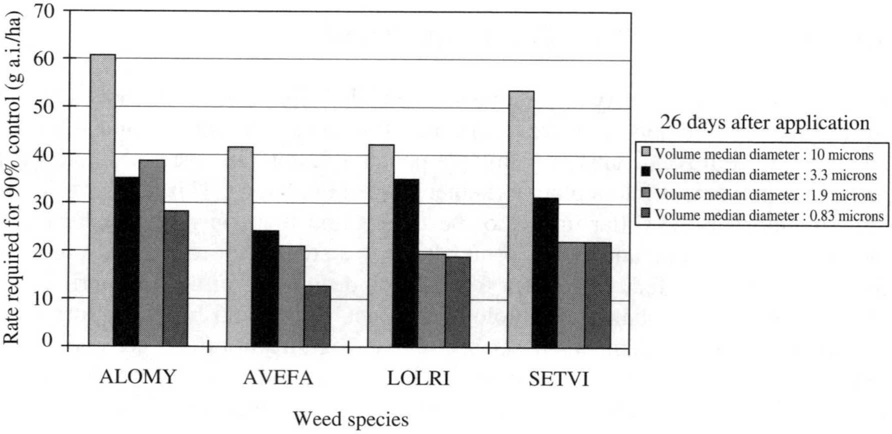

Figure 7.5 Efficacy of tralkoxydim suspension formulations vs. particle size (Silverthorne and Murfitt, pers. comm.)

Table 7.1 Relative performance of 'Reflex' and 'Flexstar' formulations of fomesafen (University of Illinois 1996)

Weed species	'Reflex' score	'Flexstar' score
Velvetleaf	7	8
Cocklebur	8	8+
Common ragweed	8	9
Burcucumber	6	7
Eastern black nightshade	7+	8
Kochia	5	6
Wild sunflower	7	8

Where: 9 = 85 to 95% control, 8 = 75 to 85% control etc., + indicates control at the top end of the range.

Table 7.2 Effect of formulation type on flurochloridone phytotoxicity to sunflowers (Beraud, pers. comm.)

Formulation	Rate (g a.i./ha)	% Plants showing bleaching	% Leaf surface bleached	% Crop vigour
'Racer' CS	750	1.4	2.5	100
	1500	10	9.6	91
'Racer' EC	750	23.1	9.8	99
	1500	47	27.7	69.5

Another associated aspect in enhancing the effectiveness of the applied pesticide involves the degree to which the crop might be harmed by the pesticide application. Much literature has been generated on minimising the crop damage by the careful selection of formulation type and formulation components. One particular interest area has been the importance of solvent selection (Krenek and King, 1997). Another theme has been the choice of adjuvants and how this might influence crop damage. This is of particular interest to formulators of fungicide products, where fungal infections need to be destroyed without harming the crop (Steurbaut *et al.*, 1989). Furthermore, the choice of formulation type itself can be influential. An example of the influence of formulation type is provided by the case of flurochloridone. This pre-emergence herbicide can be prepared as either an emulsifiable concentrate (EC) or as a microcapsule (CS) formulation. Table 7.2 shows the difference in the degree to which the formulation type influences the crop damage caused to sunflowers by flurochloridone.

CONCLUSION

The above sections show that an intelligent and scientific approach to formulation of pesticides can bring about significant improvements to the efficiency of their use. It should also be apparent that, although much progress has been made, there is still much scope for further improvements. That microencapsulation technology has been so prominent in many of the improvements seen to date should not come as a surprise, as this technology provides the formulator with a unique tool for manipulating the delivery of the pesticide. This area of technology, perhaps better defined as triggered and controlled release, will continue to play an important part in helping formulators deliver further improvements to the product offer. Pesticide formulators will continue to play a significant role in developing future improved pesticide products.

From the formulator's perspective, the onward development of current technologies will lead to a steady improvement in the effectiveness of pesticide products. More significant, or step-change, improvements will require a significant change in the way in which pesticides are used. Whether this is provided entirely by external changes (e.g. biotechnology) or by an internal change (e.g. the reintroduction of electrostatic spraying technology) has yet to be seen.

ACKNOWLEDGEMENTS

The author wishes to acknowledge the help and advice provided by Dr Günter Menschel of the BBA, Germany, Dr Herb Scher, formerly of Syngenta, Richmond, California, and John Endacott and many other colleagues within Syngenta.

'Spraymate Bond' is a registered trademark of Newbrook Agricultural Products, UK.

'Nufilm P' is a registered trademark of Miller Chemical & Fertiliser Corporation, USA.

'Reflex', 'Flexstar' and 'Racer' are registered trademarks of Syngenta.

REFERENCES

Anderson NH and Hall DJ The role of dynamic surface tension in the retention of surfactant on pea plants, *Adjuvants and Agrochemicals*, Volume 2: Recent Development, Application, and Bibliography of Agroadjuvants, pp. 51–62, Chow, P.N.P., Grant, C., Hinshalwood, A.M. and Simundsson, E. (eds) (1989).

Anderson NH, Hall DJ and Seaman D; Spray retention: effects of surfactants and plant species, *Aspects of Applied Biology*, **14**: 233–243 (1987).

Anon., *Achtergronddocument MJP-G Emissie Evaluatie*, IKC Ede, The Netherlands 1996.

Backman PA, Munger GD and Marks F; The effects of particle size and distribution on performance of the fungicide chlorothalonil, *Disease Control, Phytopathology*, **66**(10): 1242–1245 (1976).

Baylis AD and Hart CA; Varying responses among weed species to glyphosate-trimesium in the presence of an organosilicone surfactant, '*Proceedings, 1980 British Crop Protection Conference – Weeds*', pp. 1331–1336 (1993).

Beraud JM; Syngenta Agrochemicals, unpublished data (2000).

Bohannen DR and Jordan TN; Effects of ultra-low volume applications on herbicide efficacy using oil diluents as carriers, *Weed Technology*, **9**: 682–688 (1995).

Burnett, Michael, Syngenta Agrochemicals, Jealots Hill Research Station, UK.
Chapple AC, Downer RA and Hall FR; Effects of spray adjuvants on swath patterns and droplet spectra for a flat-fan hydraulic nozzle, *Crop Protection*, **12**: 579–590 (1993).
Coffee RA; Electrodynamic crop spraying, *Outlook on Agriculture*, **1**: 350–356 (1981).
Curtis R, Jain R, Creech D and Fitch W; *US Patent 5 462 915* (1995).
Dexter RW and Huddleston EW; Effects of adjuvants and dynamic surface tension on spray properties under simulated aerial conditions, *Pesticide Formulations and Application Systems: Eighteenth Volume, ASTM STP 1347*, Nalewaja JD, Goss GR and Tann RS (Eds), American Society for Testing and Materials (1998).
Ebert TA, Taylor RAJ, Downer, RA and Hall F; Deposit structure and efficacy of pesticide application. 1: Interactions between deposit size, toxicant concentration and deposit number, *Pesticide Science*, **55**: 783–792 (1999).
Elliott JG and Wilson BJ (Eds); The Influence of Weather on the Efficiency and Safety of Pesticide Application: the Drift of Herbicides. *British Crop Protection Council Occasional Publication No. 3. Report of the Working Party of the BCPC Research Development Committee.*
Field RJ and Bishop NG; Promotion of stomatal infiltration of glyphosate by an organosilicone surfactant reduces the critical rainfall period, *Pesticide Science*, **24**: 55–62 (1988).
Ganzelmeier, H; *Plant Protection – Current State of Technique and Innovations, Pesticide Chemistry and Bioscience (The Food – Environment challenge)*; Brooks GT and Roberts TR (eds). RSC (Royal Society of Chemistry), UK (1999).
Gaskin RE and Stevens, PJG; Antagonism of the foliar uptake of glyphosate into grasses by organosilicones surfactants. Parts 1 and 2: Effect of plant species, formulation, concentrations and timing of application. *Pesticide Science*, **38**: 185–200 (1993).
Groenwald BE, Pereiro F, Purnell TJ and Scher HB; Microencapsulated thiocarbamate herbicides: a review of their physical, chemical and biological properties, *Proceedings, British Crop Protection Conference – Weeds*, pp. 185–191 (1980).
Hanks JE; Herbicide use reduction by improved application technology. In: *Clean water – Clean Environment – 21st Century: Team Agriculture – Working to Protect Water Resources.* American Society of Agricultural Engineers (ASAE); Conference proceedings Volume 1: Pesticides. March 5–8, 1995, Kansas City, Missouri. (1996).
Hart CA and Young BW, Scanning electron microscopy and cathodoluminescence in the study of interactions between spray droplets and leaf surfaces, *Aspects of Applied Biology*, **14**: 127–140 (1987).
Hazen JL; *US Patent 5550224* (1996).
Humphrey ST; Agrochemical formulations using natural lignin products, In: *Chemistry and Technology of Agrochemical Formulations*, Knowles DA (ed.). Kluwer Academic Publishers, The Netherlands (1998).
Knoche M and Bukovac MJ; Interaction of surfactant and leaf surface in glyphosate adsorption, *Weed Science*, **41**: 87–93 (1993).
Krenek MR and King DN; The relative phytoxicity of selected hydrocarbon and oxygenated solvents and oils, *Pesticide Formulations and Application Systems: 6th symposium*, STP 943 ASTM, Philadelphia, PA, USA (1997).
Kudsk P; The effect of adjuvants on the rainfastness of thifensulfuron and tribenuron, In: *Adjuvants for Agrochemicals*, Foy CL (ed.), Wiley Interscience, UK (1992).
Kudsk P and Mathiassen SK; Influence of formulations and adjuvants on the rainfastness of maneb and mancozeb on pea and potato, *Pesticide Science*, **33**: 57–71 (1991).
Lee FTH and Nicholson P; *International Patent: WO 96/14743.*
Maas W; Influence of particle size on pesticidal activity, *Advances in Pesticide Science*, Geissbuehler H (ed.), Volume 3, pp. 772–779 (1979).
Matsumoto S, Suzuki H, Tomita H and Shigematsu T; Effect of humectants on pesticide uptake through plant leaf surfaces, In: *Adjuvants for Agrochemicals* Foy CL (ed.), CRC Press, Boca Raton, Florida, USA (1992).

Nalewaja JD and Matysiak R; Salt antagonism of glyphosate, *Weed Science*, **39**: 622–628 (1991).

Nalewaja JD and Matysiak R; Spray carrier salts affect herbicide toxicity to kochia (*Kochia scoparia*), *Weed Technology* **7**(1): 154–8 (1993).

Nalewaja JD, Matysiak R and Szeleniak E; Sethoxydim response to spray carrier chemical properties and environment, *Weed Technology* **8**(3): 591–597 (1994).

Paterson ES; *IUPAC Poster* (1998).

Reekmans S, Holloway PJ, Davies SJ and Blease T; Structure activity correlations for distinct adjuvant chemistries. *IUPAC Poster* (1998).

Schaefer GW and Allsop K; Spray droplet behaviour above and within the crop, *Proceedings of the 1983 British Crop Protection Conference*, Brighton, UK, pp. 1057–1065 (1983).

Scher HB, Rodson M and Lee K; Microencapsulation of pesticides by interfacial polymerisation utilising isocyanate or aminoplast chemistry, *Pesticide Science*, **54**: 394–400 (1998).

Schönherr J, Baur P and Buchholz A; Modelling foliar penetration: its role in optimising pesticide delivery, *Pesticide Chemistry and Bioscience (The Food – Environment Challenge)*; Brooks GT and Roberts TR (Eds), Royal Society of Chemistry, Cambridge (1999).

Silverthorne J and Murfitt R; Syngenta Agrochemicals, unpublished data (2000).

Spray Drift Taskforce *A Summary of Spray Application Studies*, Stewart Agricultural Research Services, Missouri (1997).

Steurbaut W, Megahed HS, Van Roey G, Melkebeke T and Dejonckheere W; Improvement of fungicide performance by the addition of surfactants to the formulations, Part II Influence on biological and systemic activity, *Med. Fac. Landbouww. Rijksuniv. Gent* **54**: 219–232 (1989).

Stevens PJG, Gaskin RE, Hong SO and Zabkiewicz JA; Contributions of stomatal infiltration and cuticular penetration to enhancements of foliar uptake by surfactants. *Pesticide Science*, **33**: 371–382 (1991).

Stock D, Edgerton BM, Gaskin RE and Holloway PJ; Surfactant-enhanced foliar uptake of some organic compounds: interactions with two model polyoxyethylene aliphatic alcohols, *Pesticide Science*, **34**: 233–242 (1992).

Thelen KD, Jackson EP and Penner D; The basis for hard water antagonism of glyphosate activity, *Weed Science*, **43**: 541–548 (1995).

University of Illinois, *Agricultural Pest Management Handbook* (1996).

US Environmental Protection Agency (USA) website provides a good guide, e.g. http://www.epa.gov.

Westvaco Technical Brochure (2000) *Kraft Lignins and Sunlight Sensitive Compounds*.

Wirth W, Storp S and Jacobsen W; Mechanisms controlling leaf retention of agricultural spray solutions, *Pesticide Science*, **33**: 411–420 (1991).

8 Rational Pesticide Use: Spatially and Temporally Targeted Application of Specific Products

R. BATEMAN
IPARC and CABI Bioscience, Ascot, Berks UK

INTRODUCTION

Attitudes to pest management became polarised over the final three decades of the 20th century. Prior to this, the 1950s and 1960s, pesticides were seen by many as a panacea for pest problems. However, there was dissension, expressed most famously by Carson (1962). There followed a reappraisal of the role of pesticides, and a 'rediscovery' of the importance of biological control mechanisms (the manipulation of which had its roots in the early part of the century). Not surprisingly perhaps, pesticide companies were slow to accommodate the Integrated Pest Management (IPM) concepts that developed in the 1970s, and the industry was perceived to be operating a conspiracy (van den Bosch, 1980). In brief, the major criticisms of pesticides are that their use may:

- cause poisoning to humans (e.g. Bull, 1982; Loevinsohn, 1987);
- cause environmental damage (Schnoor, 1992);
- be costly to farmers, growers or governments (Bull, 1982);
- create a 'pesticide treadmill', leading to cases of pesticide-induced resurgence of pests that would otherwise be of minor importance (van den Bosch, 1980); and
- be unsustainable, with the development of pest resistance (e.g. Georghiou, 1986)

Several authors, including van Emden and Peakall (1996), have pointed out that pesticides are important tools and have saved many millions of lives in malaria control as well as economic crop protection. They suggest that the environmental impact of pesticides has often been overstated, with pollutants and other

Optimising Pesticide Use Edited by M. Wilson
 ISBN: 0-471-49075-X

human activities (including the conversion of natural habitats into agricultural land) causing much greater damage. Even naturally occurring substances (notably micro-organisms and their metabolites) may present health risks (Graham Bryce, 1989). However, the undesired effects – especially of organochlorines on agricultural environments – have been real and costly; on the other hand, the problem of thinning raptor eggshells has been substantially solved, by the banning of organochlorine insecticides (Newton, 1995). van Emden and Peakall (1996) have discussed the economic assessment of hazards to human health and the environment, including those arising from industrial processes associated with their manufacture.

In the second half of the 1980s, an IPM paradigm emerged that placed heavy emphasis on large-scale, participatory training networks, enabling farmers to understand the ecological interactions in their crops. With World Bank loans, bilateral and multilateral aid, many millions of US$ were spent on IPM programmes: notably in SE Asia, where pesticide inputs on rice could be dramatically curtailed, to the economic benefit of smallholder farmers (Matteson, 2000). More judicious (or no) use of insecticides reduced the risk of resurgence of pests such as planthoppers and leafrollers. This was brought about by raising the level of awareness of the interactions between the crop, pests, pesticides and natural enemies in farmer field schools (Röling and van de Fliert, *in* Röling and Wagemakers, 1998). However, concerns have been expressed about the sustainability of these programmes when external funding comes to an end, so they become 'institutionalised'. Similar programmes have achieved localised success in other parts of the world (Röling and Wagemakers, 1998), but, in reality, upward trends in pesticide use have continued in most countries during the 1990s.

Even the most high-profile publicly funded programmes cannot match the resources put into promotion of pesticide products; some economists would also point here to a long-term failure of aid in comparison with true markets (regulated in the case of pesticides). I will not attempt to enlarge on these politically charged and complicated issues, but I intend to identify some of the *current real problems* in pesticide use and how they might be addressed.

The global value of pesticide sales has continued to rise: by a factor of 2.5 times over the last 20 years (Figure 8.1), although the rate of increase in real (weighted) terms has declined from the 5% reported by Bull (1982) to an average of 1.1% over the last five years. It would have been more useful to express sales in terms of hectare dosages, but such information is only rarely available (http://www.inet.uni2.dk/~iaotb/top20.htm is an exceptional database on Danish usage). There have now been three major downturns: in 1982–1983 and during the beginning and end of the 1990s; these have been attributed primarily to recessions and falls in agricultural commodity prices. Essentially though, these figures probably reveal that the pesticide market has 'matured' rather than having declined.

Within the overall pesticide market, the proportion of insecticide sales decreased during the 1980s (from about 32% to 29%), but has changed little over the last

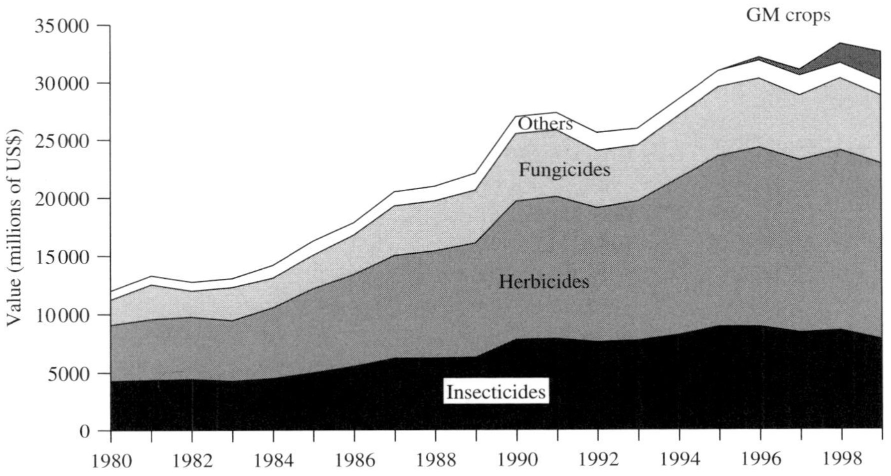

Figure 8.1 Worldwide pesticide markets in the final two decades of the 20th century. Data compiled from annual reviews of the Crop Protection Association

decade. The fungicide sector (which includes seed treatments) has only fluctuated slightly around the 20% mark over the whole period, but this proportion may rise if conventional insecticide markets decrease. The most noticeable increase has been in herbicide sales: from 40% in 1980 to 48% in recent years. The introduction of genetically modified (GM) crops over the last four years may influence the pesticide market, but it is too early to predict the progress and repercussions of this controversial sector. For example, Gianessi and Carpenter (2000) found a 9% fall in herbicide use per hectare in 13 American states growing transgenic soya beans, but yields had not changed. Paradoxically though, because of an increased area of cultivation (possibly linked to improved work rates), total herbicide use in these states increased by 14%.

IPM is a strategy that aims to optimise all available techniques, in order to maintain pest populations at levels below those causing economic injury. Although many believe that IPM may result in reduction or removal of chemical pesticide use, in practice, farmers may need pesticides to prevent devastating losses in certain crops. Clearly, arguments that pesticide use should be abandoned are unrealistic (especially with nonstaple crops). Rational pesticide use (RPU) was coined in the title of a book by Brent and Atkin (1987); it can be defined as a focused subset of IPM, in which the adverse effects of pesticide use can be mitigated by improvements in the selectivity of the products themselves and the precision of their application in both space and time. A more rigorous approach to application techniques can also improve operator safety, resistance management and, perhaps of most interest to the farmer, reduce pesticide costs. Similar concepts have been promoted elsewhere, which all attempt to achieve

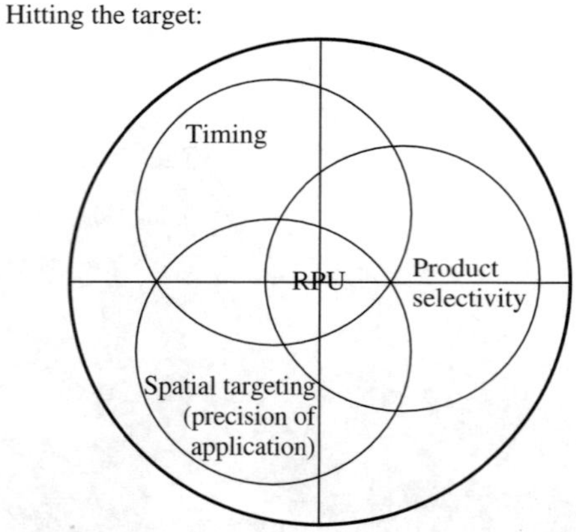

Figure 8.2 Principal elements of rational pesticide use

sustainable agriculture, with a low environmental impact, achieved through a combination of appropriate technologies, but not necessarily excluding the use of pesticides. Examples include: *lutte raisonée* (supervised control) promoted by the FARRE project in France (Milaire, 1995); 'green agriculture', promoted by the Chinese Academy of Agricultural Sciences (which, together with 'white agriculture' – i.e. biotechnology – focuses especially on techniques such as the use of microbial agents); and 'clean production' in Vietnam (where crop protection may include limited use of pesticides, and is combined with more careful use of human waste as a fertiliser).

Having accepted that there is a need for pesticides and that their use will continue for the foreseeable future, RPU focuses on the technologies that minimise the costs and deleterious side effects of such interventions. I shall start by dealing with its three major elements separately; however, it is where they can be combined (Figure 8.2) that the greatest benefits accrue.

SELECTIVITY OF PRODUCTS

The most common approach taken to reduce the impact of pesticides on non-target organisms has been the development of selective active ingredients (a.i.s: both biological and chemical). Although I will argue that reliance on this tactic alone is a mistake, it is an obvious starting point.

THE CHANGING NATURE OF CHEMICAL PESTICIDES

Table 8.1 lists some of the currently most widely used pesticides, together with selected historically important and technically significant examples (Tomlin, 2000,

Table 8.1 Some characteristics of chemical and biological pesticides

Pesticide and date of introduction	Typical application rate (g a.i/ha)	Toxicity class (WHO/EPA)	Spectrum
Insecticides			
DDT (*organochlorine*), 1942	1000–2500	II	Broad
*Endosulfan (*organochlorine*), 1956	500–750	I–II	Moderate
*Chlorpyrifos-ethyl (*organophosphorus*), 1965	500–750	II	Broad, degradation slower than other OPs
*Monocrotophos (*organophosphorus*), 1965	150–1500	I	Broad–narrow[5] (S)
Carbaryl (*carbamate*), 1957	250–2000	II–III	Broad
*Carbofuran (*carbamate*), 1965	500–750	I (SC); II (GR)	Broad (S)
*Cypermethrin (*pyrethroid*), 1975	25–100	II	Broad
Deltamethrin (*pyrethroid*), 1974	5–15	II	Broad
Diflubenzuron (*benzoylurea*), 1972	25–100	III	Narrow (larvae)
Fipronil (*phenylpyrazole*), 1987	5–80	II–III	Broad
*Imidochloprid (*neo-nicotinoid*), 1990	10–100 (sprays) 0.5–7 g/kg seed	II–III	Broad (sprays) (S) Narrow (seeds)
Spinosad (*fermentation product*), 1992	5–200	IV	Moderate
Pymetrozine (*azomethine*), 1992	150	III	Selective (S) (Homoptera)
Biological insecticides: Bacillus thuringiensis kurstakii, 1902 (products: 1938)	10–40 (5–20 × 10^9 IU)	III	Narrow
Metarhizium anisopliae: species 1878; products 1974, 1990 (var. *acridum*) 1999	40–100 (2–5 × 10^{12} conidia)	U	Specific
Spodoptera exigua nucleopolyhedrovirus, 1986	Approximately 10^{12} polyhedra	U	Highly specific
Multifunctional agents			
Petroleum oils (fungicide, acaricide, insecticide, adjuvant), *c.*1922 – but a range of modern products now available	1–22.5 l/ha	IV	Moderate
*Sulfur (fungicide, acaricide), Ancient	1000–20 000	IV	Moderate

(*continued overleaf*)

Table 8.1 (*continued*)

Pesticide and date of introduction	Typical application rate (g a.i/ha)	Toxicity class (WHO/EPA)	Spectrum
Fungicides			
**Copper fungicides*: $Cu(OH)_2$, $Cu_2Cl(OH)_3$ $CuSO_4 \cdot 5H_2O$, 19th century	1000–7500 (often mixtures)	I–III	Broad-spectrum, not degraded in soil
*Mancozeb (*dithiocarbamate*), 1961	1000–3600	IV	Moderate, rapidly degraded
*Chlorothalonil (*substituted aromatic*), 1964	500–1500	III–IV	Broad-spectrum
*Metalaxyl (*phenylamide*), 1977	500–1500	III	Broad-spectrum (S)
Azoxystrobin (*strobilurin*), 1992	100–375	III–IV	Broad-spectrum (S-)
Benomyl (*benzimidazole*), 1968	14–1100	III–IV	Broad-spectrum (S)
*Propiconazole (*triazole*), 1979	100–150	III	Broad-spectrum (S)
Ampelomyces quisqualis (*hyperparasitic mycofungicide*), isolated: 1984, product: 1995	$1.8–3.5 \times 10^{11}$ (20–40g)	IV	Specific to powdery mildews
Herbicides			
*2,4, D (*aryloxyalkanoic acid*), 1942	280–2300	II	Dicotyledons (S)
*Atrazine (*triazine*), 1957	100–2000	III	Dicotyledons (S)
*Paraquat dichloride (*bipyridylium*), 1958	500–2500	II	Broad-spectrum
*Glyphosate-isopropylammonium, 1971 ~ trimesium,• 1989 (*OPs*)	250–2000	III	Broad-spectrum (S)
*Imazethapyr (*imidazolinone*), 1984	50–100	III	Broad-spectrum (S)
*Metolachlor (pre-emergence *chloroacetanilide*), 1974	1000–2500	III	Narrow (S)
Bensulfuron-methyl (*sulfonylurea),* 1983	20–75	IV	Selective: especially rice (S)
Colletotrichum gloeosporioides f. p. Aeschynomene (mycoherbicide), 1985	1.71×10^{11} conidia	U	Specific to *Aeschynomene virginica*

Note: '*' indicates important pesticides at the time of writing (top five in class). (S) indicates systemic activity.

and manufacturers' literature). Several of the compounds still in widespread use were discovered during the early days of pesticide development; their development patents have expired and they are known as 'generics'. Examples include the insecticides endosulfan (patented in 1956), chlorpyrifos-ethyl (1963) and cypermethrin (1975); fungicides such as copper compounds (19th century) and mancozeb (1961); herbicides such as: 2,4-D (1942), atrazine (1957), and multi-functional compounds such as sulfur and petroleum oils. The latter are staging a 'comeback' after several decades, as improved, purified, proprietary products, due to heightened interest in IPM and organic production methods. Likewise, there is interest (although only a small market at present) in biopesticides, which will be discussed in the next Section.

Field dosages often span a very broad range, and the 'typical application rates', shown in Table 8.1, are for general comparison only. Nevertheless, these numbers indicate how field a.i. rates have decreased historically, as more active molecules have been developed; examples include the neo-nicotinoid insecticides, triazole fungicides and sulfonylurea herbicides, which are one to two orders of magnitude more active than the early organochlorine, copper and aryloxyalkanoic acid pesticides, respectively.

Mammalian toxicity information has been summarised as LD_{50} values for skin and eye toxicity (usually the most important factor for spray operators) and typical WHO/EPA toxicity classes (an overall measure of risk for formulated products). 'Pesticide poisoning' is an expression that is used loosely; by far the greatest danger is with insecticides. Mammalian toxicity risk is generally much less of an issue with fungicides and most herbicides (paraquat being a notable exception), even in comparison with everyday substances such as aspirin and caffeine (as pointed out by Graham Bryce in 1989). High numbers of human casualties have been caused by OPs and the cyclodienes such as aldrin, endrin and endosulfan. Although lower-toxicity OPs (e.g. malathion), certain carbamates (e.g. carbaryl) and many of the acaricides had been available for a long time, even by the 1980s it was only in a minority of countries that the hazard to humans was taken seriously by restricting access to the more risky substances (Bull, 1982).

Over the last three decades, pesticide companies have, on the whole, developed less-toxic insecticides. Unfortunately, during this period, specific compounds (e.g. the insect growth regulators, biopesticides and fermentation products such as spinosad) have tended to only fill 'niches' in the pesticide market. With huge development costs, companies are put under great pressure to develop those pesticides which have a broad spectrum and thus a wide market. Neo-nicotinoids are a good example; since Bayer introduced imidacloprid in 1990, it has rapidly become the most profitable insecticide on the market, and other major companies are in the process of developing their own rival molecules. A similar explosion of interest in pyrethroids occurred during the early 1980s, and they remain very popular with farmers. Both these groups have broad spectra, and more specific compounds (even aphicides) by definition occupy small proportions of the pesticide market and therefore a smaller potential return on research investment.

Paradoxically, therefore, it can be argued that one of the costs of heightened regulatory pressure has been an increase in the impact of pesticides on nontarget organisms and the environment, since the OPs were prevalent. The practical – quite apart from the environmental – disadvantages of impact on beneficial organisms include pesticide induced resurgence of relatively unimportant insects and mites. Among the early observations of this phenomenon were outbreaks of spider mites (*Panonychus ulmi*) in top fruit (Way and van Emden, 2000), and brown planthoppers (*Nilaparvata lugens*) in tropical rice (Heinrichs, 1979). The risk of resurgence in rice is so high that it may be most straightforward to discourage insecticide use altogether (Matteson, 2000). However, the use of herbicides (for directly seeded rice) and fungicides (against diseases such as sheath blight) is increasing in some countries, and considerably more complex solutions appear to be necessary in other crops. The need for more specific insecticides is often identified in farmer field schools and by scientists, but, unfortunately, not always by the marketplace. However, there has been recent large-scale development of new chemical groups with specific action, provided that the market is sufficiently important; an example is pymetrozine, which has a totally novel mode of action against aphids and whiteflies (Kayser *et al.*, 1994).

The key issues with fungicides are growing concerns about residues and resistance (especially the benzimidazoles, dicarboximides, phenylamides and sterol biosynthesis inhibitors or SBIs). Fungicide resistance-management strategies are based on restricting and rotating the use of chemical groups; mixtures of a.i. are also becoming common. Newer fungicides cover a very diverse range of chemical groups (Gullino *et al.*, 2000); recently, the most commercially important of these have been the strobilurin fungicides, which, being fermentation products, have a more 'biorational' image than some of the older chemicals.

Herbicide resistance is currently less of a problem than with the other pesticide groups, although, once it occurs, the effects are profound. Instead, the major interest centres around the control of off-target drift (i.e. application issues), specificity and the use (and side-effects) of certain herbicides with genetically modified crops. The examples listed in Table 8.1 indicate the continuing importance of pre-1980 molecules. For instance, glyphosate has become the most important herbicide, with its recent fall in price, broad spectrum of efficacy, low mammalian toxicity and (where GM crops are acceptable) its compatibility with 'Roundup ready' crops.

Generic pesticides (those that have exceeded their patent life) represent over half of the global crop protection market (US$18 000 million of end-user sales in 1998). The patent expiry of important molecules such as the pyrethroids, triazole fungicides and glyphosate has resulted in substantially reduced prices, and there have been tensions between the major multinational and generic (often national) pesticide companies. Nevertheless, some 70% of the profits from multinational companies come from compounds that have exceeded their normal, 20-year patent life (i.e. approximately 13 years of profitable life: Finney, 1988). In response, the major multinational companies have attempted to maintain their market share by increased 'product stewardship', and it is the proprietary nature of product databases (including toxicology, efficacy and resistance management) that has become crucial. Other approaches have included the promotion of 'integrated crop management' (ICM) packages to the farmer (that includes the development and provision of genetically modified seeds and pesticide mixtures). 'Added value' can also be conferred to off-patent pesticides by developing new formulations and manufacturing processes; Carroll (1999) describes how purified or single isomer products can 'rejuvenate' older molecules.

Over the last 20 years, there have also been substantial improvements to the formulation that have enhanced selectivity and improved operator safety (Mulqueen, 1998). These include suspension concentrates (SC), capsule suspension (CS), and other formulations which are sprayed as particulate suspensions; in this way certain aspects of chemical application may have become similar to those of biopesticides (e.g. as in Figure 8.8). Innovations with biopesticide formulation have sometimes proved momentous: for example, the discovery that mycoinsecticide efficacy could be substantially enhanced by formulating in oil (Prior *et al.*, 1988).

BIOPESTICIDES AND OTHER SELECTIVE ACTIVE INGREDIENTS

Copping (1998) provides an overview of the currently available biopesticide products, which have been promoted for a long time, but their use remains limited to <1% of the total pesticide market. Lisansky (1997) attributes this limited growth to lack of positive promotion by the authorities and the agrochemical industry. Their perceived constraints include: narrow target spectra, poor performance relative to cost and inconsistent product quality in comparison with chemicals. It is clear that biopesticidal agents will only become successful on the basis of demonstrable advantages, and will not necessarily be accepted simply because they are of natural origin.

O'Connell and Zoschke (1996) give an industry perspective on the relatively disappointing progress to date in the development of biological herbicides. Like several other authors, they stress the need for active agents, identification of suitable target (weed) pests, and the need for interdisciplinary technical cooperation to overcome technical hurdles and appropriate registration procedures. The mycoherbicide example shown in Table 8.1 is 'Collego', which was technically successful for the control of northern jointvetch (*Aeschynomene virginica*: Leguminosae) in North American rice and soyabeans. 'Collego' is still available, but commercial sales have been limited. Review of the scientific literature may give a misleading impression of the true status of biopesticides, with authors more inclined to record successes rather than failures or discontinued products.

Mycofungicides are still at an early stage of development, but perhaps have better commercial prospects in field crops. For example, there is a clear need to reduce the number of applications of sulfur (used only before crop flowering) and two to three final sprays of SBI fungicides to vines; *Ampelomyces quisqualis* is specific to powdery mildews (several genera in the Erysiphales), and can be used in IPM programmes (Hofstein and Chapple, 1999). Harman (2000) has shown that products based on a *Trichoderma harzianum* isolate provide effective and reliable alternatives to chemical fungicides: partly through phytotonic action.

The market for biological insecticides remains a promising sector, with one bacterial agent (*Bacillus thuringiensis*) firmly established as a number of products. Copping (1998) lists a number of other microbial agents, based on viruses, nematodes and fungi, which have been the subject of special interest because of the particular concerns associated with chemical insecticides. By monitoring efficacy over weeks (rather than days), some biopesticides are substantially more efficacious than their chemical rivals. This can be illustrated by results gained with the international LUBILOSA[1] Programme, which is dedicated to the replacement of chemical pesticides with environmentally friendly alternatives such as Mitosporic (deuteromycete) fungi. Extensive locust spray campaigns in the 1980s caused public concern about the accompanying environmental impact of chemicals, and biological pest management methods are especially important for pest control in natural and semi-natural habitats.

The Programme has developed a mycoinsecticide ('Green Muscle') based on an isolate of *Metarhizium anisopliae* var. *acridum*, which has been field tested against a number of acridid pest species (Lomer *et al.*, 1999), and is now recognised as an appropriate product for locust control in environmentally sensitive areas (FAO, 1997). In a recent series of operational trials against *Oedaleus senegalensis*, it was shown that although the organophosphorus chemical fenitrothion achieves an impressive 'knock down', hopper populations recover two weeks after application, and a more profound population reduction was achieved in the plots sprayed at ultra-low volume (ULV) rates with *Metarhizium* conidia (Figure 8.3; Langewald *et al.*, 1999). Residues of the fungus are more persistent, but have minimal impact on nontarget organisms – which may also enhance field efficacy (Peveling *et al.*, 1999). After the death of the insects over 4–20 days, further conidia are produced and slowly released as the cadavers are broken down; under suitable conditions secondary cycling of inoculum may also reduce acridid populations (Thomas *et al.*, 1995). They therefore point out that simply using a 'chemical model' to describe the selective performance of biopesticides seriously undervalues their full potential. Nevertheless, in the example shown (Figure 8.3), the perception of efficacy (by the third week after application) cannot be due to secondary cycling, because there had not been time for this process to occur. Of all the modes of dose transfer, secondary pick-up of spray residues is often the most important (e.g. Bateman *et al.*, 1998), thus application of *Metarhizium* is not unlike that of other (chemical) products.

Effective extension is crucial. The LUBILOSA Programme managed to create a market for 'Green Muscle', at the same time as demonstrating its efficacy, with

[1]LUtte BIologique contre les LOcustes et les SAuteriaux: a collaborative, multidisciplinary research and development programme funded by the Governments of Canada (CIDA), The Netherlands (DGIS), Switzerland (SDC) and the UK (DfID).

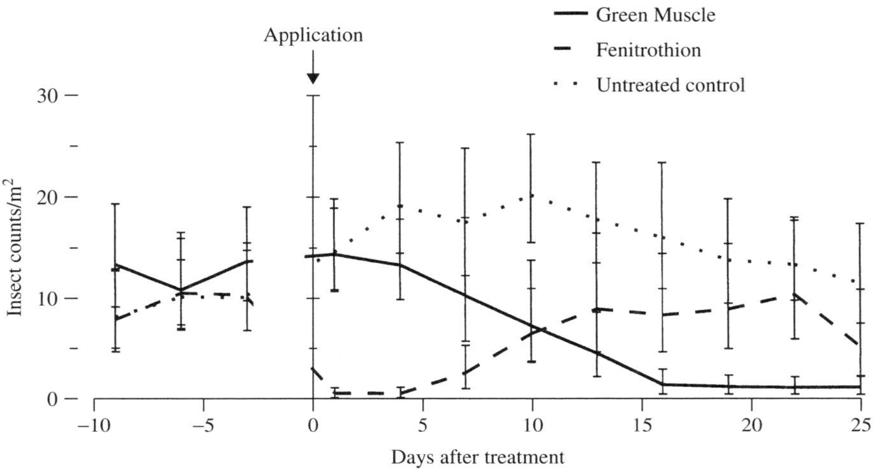

Figure 8.3 Results of an 800 ha aerial spray trial against *Oedaleus senegalensis*, carried out in 1997 at Maine Soroa, eastern Niger (from Langewald *et al.*, 1999). There were three pre-treatment sampling days; bars are ± s.e. on means for back-transformed population counts

an extensive network of farmer participatory training and research in a number of Sahelian countries. Of all the biopesticides, *B. thuringiensis* (*B. t.*) remains by far the most well known and important, with products are available in most markets. As part of the South and South-East Asia intercountry programme for IPM in vegetable growing, another international, collaborative team has sought to promote the use of *B. t.* as an alternative to broad spectrum, toxic chemicals for caterpillar control (Jenkins and Vos, 2000). However, it was found necessary to develop at least four exercises in order for farmers to appreciate the key aspects of effective *B. t.* use:

1. a comparison of the impact of *B. t.* and a toxic standard on the target pest and its natural enemies;
2. assessments of viability of *B. t.* products after storage;
3. demonstration of how *B. t.* inhibits larval feeding; and
4. sensitivity of *B. t.* deposits to sunlight.

There are interesting similarities between the promotion of biopesticides and Controlled Droplet Application (CDA) techniques (see below). As biological control agents, microbes (unlike insects) usually cannot be seen with the naked eye, and they are slow acting; therefore explaining their modes of action can be difficult. ULV application, using very fine droplets, can similarly be an 'act of faith', with farmers who are unable to see spray clouds or deposits. Both improved application technologies and biopesticides may be useful tools to solve key pest problems, but there are challenges in the availability of products, their quality and extension. Both are biology-based technical solutions waiting to be promoted and adopted more widely.

PRECISE SPATIAL APPLICATION

Over the past three decades, application technologies such as CDA have received extensive interest, but, as with biopesticides, the use of commercial products has been disappointingly limited. By controlling droplet size, ULV or very low volume (VLV) application rates of pesticidal mixtures can achieve similar (or sometimes better) biological results by improved timing and spatial application; thus CDA is a very good example of RPU technology. In contrast, conventional application using hydraulic atomisers (either on hand-held sprayers or tractor booms) with formulations mixed into high volumes of water, is known to be highly inefficient (e.g. Matthews, 1992). No atomiser has been developed that is able to produce uniform (monodisperse) droplets, but rotary (spinning disc and cage) atomisers usually produce a more uniform droplet size spectrum than conventional nozzles. Therefore CDA can be defined in terms of optimising technology to achieve a biological objective: delivering appropriately sized droplets (within practical engineering limits) for maximising the control of a given pest target (where this is known).

CDA, and the technologies that are linked to it, can be considered an example of an efficient pesticide delivery system. Optimisation of delivery systems involves not only the development of formulations and their application, but also other technical aspects of product development – at an early stage. Production and characteristics of the active ingredient (a.i.), storage and packaging must not be treated in isolation with one another, or the more obvious aspects of delivery to the biological target. The delivery-systems approach attaches equal importance to biology, technology and implementation (often effectively meaning commercial development).

Other RPU techniques include: banding, baiting, specific granule placement and weed wiping. Although useful for specific tasks, these can be labour intensive, and even baiting may have a surprisingly high, indirect environmental impact. Douthwaite and Tingle (1994) for example, describe the lower environmental impact of aerial CDA spraying for tsetse fly control in Zimbabwean woodland, in comparison with ground-based application and baiting. The baiting technique was cheap and effective, but the creation of service roads caused soil erosion and allowed poachers access to game reserves. Special emphasis will be placed here on CDA spraying, as an example of an underutilised RPU technique, with the potential to apply minimal quantities of a pesticide to maximum effect. Two other important benefits of CDA are: reduction of off-target contamination and more timely application by improving work rates.

IMPROVING THE EFFICIENCY OF ACTIVE INGREDIENTS

One of the primary aims of CDA research and development is the use of less pesticide by applying sprays in a narrow droplet spectrum. Through better *targeting* and reduction of the numbers of sprays, it may be possible to prolong the active life of new control agents before pest resistance takes place. Several laboratory and glasshouse studies with insecticides have demonstrated that smaller droplets are more efficacious for arthropod pest control than larger ones. Studies by Munthali and Wyatt (1986) investigated droplet size and concentration of oil-based dicofol formulations against immature stages of *Tetranychus urticae*; they demonstrated that median lethal doses were minimised by using the smallest possible (approximately 20 μm) droplet sizes. The relative efficacy of different droplet size spectra was reviewed by Adams *et al.* (1990); 30–60 μm droplets were usually optimal with oil-based insecticide spray deposits, but 60–120 μm was most efficient with aqueous droplets. Figure 8.4 shows that conventional hollow-cone nozzles, irrespective of cost and quality, produce broader droplet size spectra than CDA technologies. The spectra from the 'Electrodyn' and the 'Ulva+' spinning disc atomiser (illustrated in Figure 8.4) have high proportions (>80%) of spray volume in the size bands above. These and all other droplet size readings described in this chapter were measured with a Malvern 2600 particle-size analyser, using techniques described by Bateman and Alves (2000). For comparison, the illustration also includes the droplet size spectra of two hollow cone nozzles at 300 kPa pressure. 'River Mountain' nozzles (or copies of them) are arguably one of the most widely used nozzles. They are fitted to many of the hydraulic knapsack sprayers used in China and adjacent countries: 80 million

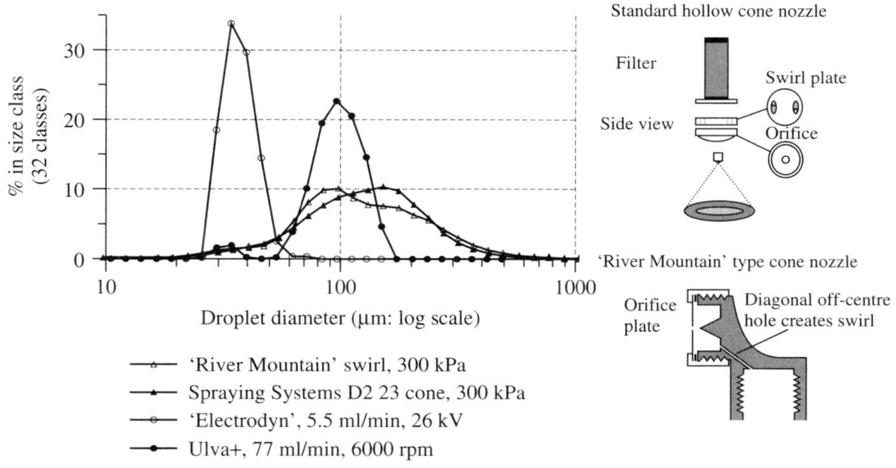

Figure 8.4 Examples of droplet size spectra: nozzles designed for insecticides and fungicides. Courtesy of Micron Ltd.

units are estimated to be in use at any given time (Chinese Academy of Agricultural Sciences, pers. comm.). Since hydraulic nozzles rely on random breakup of liquid sheets, there is little scope for narrowing droplet size spectra using this technology, and the D2 45 nozzle (more widely known in the West and much more expensive) has a similar droplet size spectrum to the 'River Mountain'.

Bals (1969) discussed the concept of producing small, uniform pesticidal droplets to achieve adequate control with 'ultra-low dosage' combined with ULV rates of application. Unfortunately, this discouraged many chemical companies from promoting CDA techniques, since sales would be reduced. An important exception occurred in the 1980s with the development of the 'Electrodyn' technology, when ICI (now Syngenta) recognised that value could be readded to lowered a.i. rates by a proprietary delivery system for smallholder farmers. By combining the bottle, formulation and nozzle in a closed, disposable system (the 'Bozzle'), operator exposure and application errors could dramatically be reduced by eliminating the mixing and measuring stages and simplifying calibration (only the forward speed needed to be monitored). Other developments in 'closed-system' techniques for packing and mixing pesticides, are described by Pfalzer (1993) and in this volume. Neither these nor the 'Bozzle' system are especially new concepts: Edward Bals had developed the Turbair ULV insecticide and 'Handy' herbicide containers by the early 1970s.

From a technical point of view, it was perhaps parallel work that made best use of previous theory and observation on the enhanced efficacy of very small droplets. Adams and Palmer (1989) used noncommercial nozzles such as the 'Microdyne' (Figure 8.5) in an airstream to produce 18 μm VMD droplets that

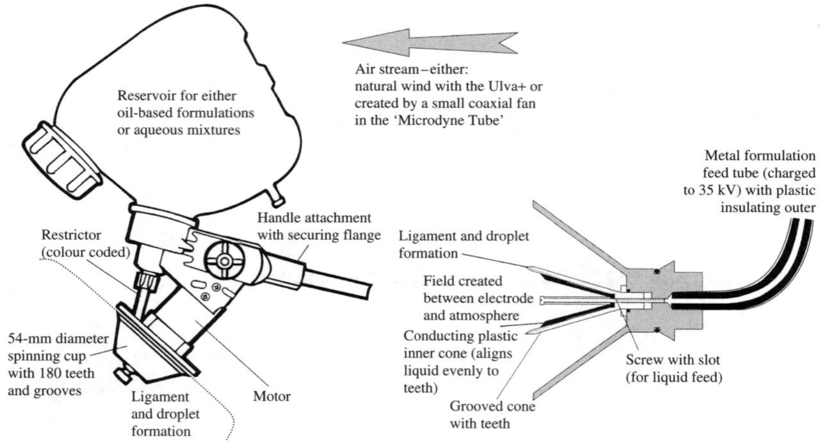

Figure 8.5 CDA atomisers (courtesy of Micron Ltd): the 'Ulva+' on the left represents the culmination of many years of small rotary atomiser development, and is robust and versatile. The 'Microdyne' atomiser (section: right) underwent much testing, but was never commercially marketed

maximised the efficiency of droplet distribution in glasshouse tomatoes. However, the critical commercial development of this research was halted because the company was not prepared to release and register appropriate formulations for 'niche' markets such as glasshouses. It is most unfortunate that 20 years after Coffee's (1979) first paper on the subject, this exciting latent technology is now no longer commercially available. Although there were technical problems (stabilising formulations, for example), these were not insurmountable for many a.i.s, and anecdotal reports put the blame on commercial and conceptual factors, including the unwillingness of farmers to commit themselves to the products of a single company.

TECHNOLOGY FOR REDUCTION OF NON-TARGET CONTAMINATION

Reducing Spray Operator Contamination

The converse of maximising pesticide contact on to the target, is the minimisation of contamination both to the operator and the environment. One of the top priorities in herbicide spraying is the accurate placement of larger (>150 μm) droplets, with minimal drift to nontarget plants. If weed leaves are the target, there is a need to balance the low-drift requirement with a droplet size that is small enough to be retained by the leaf surface. Control of droplet size also reduces the risk of operator contamination. Thornhill *et al.* (1995) showed that herbicide application with governed spinning disc sprayers (the 'Herbi' and 'Herbaflex') could reduce operator exposure by 2.7–4.4 times in comparison with lever-operated knapsack sprayers, fitted with hollow cone or deflector nozzles (Figure 8.6). However, exposure can also be reduced by training operators to improve application techniques and make minor modifications to conventional knapsack sprayers (by incorporation of devices such as constant-pressure valves).

Combining Better Targeting with Drift Reduction

With placement (localised) spraying of broad-spectrum or toxic chemicals, wind drift must be minimised, and considerable efforts have been made recently to quantify and control spray drift from hydraulic nozzles. On the other hand, wind drift is also an efficient mechanism for moving droplets of an appropriate size range to their targets over a wide area with ULV spraying. Himel (1974) made a distinction between *exo*-drift (the transfer of spray out of the target area) and *endo*-drift, where the a.i. in droplets falls into the target area, but does not reach the biological target. Endo-drift is volumetrically more significant, and may therefore cause greater ecological contamination (e.g. where chemical pesticides pollute groundwater). Unfortunately, changes in policy and practice to limit pesticide drift (especially in Europe and North America) effectively encourage the use of larger droplet sizes in ordinary spray nozzles. In other words, there is a danger of 'exodrift' being reduced at the expense of an increase in 'endodrift'.

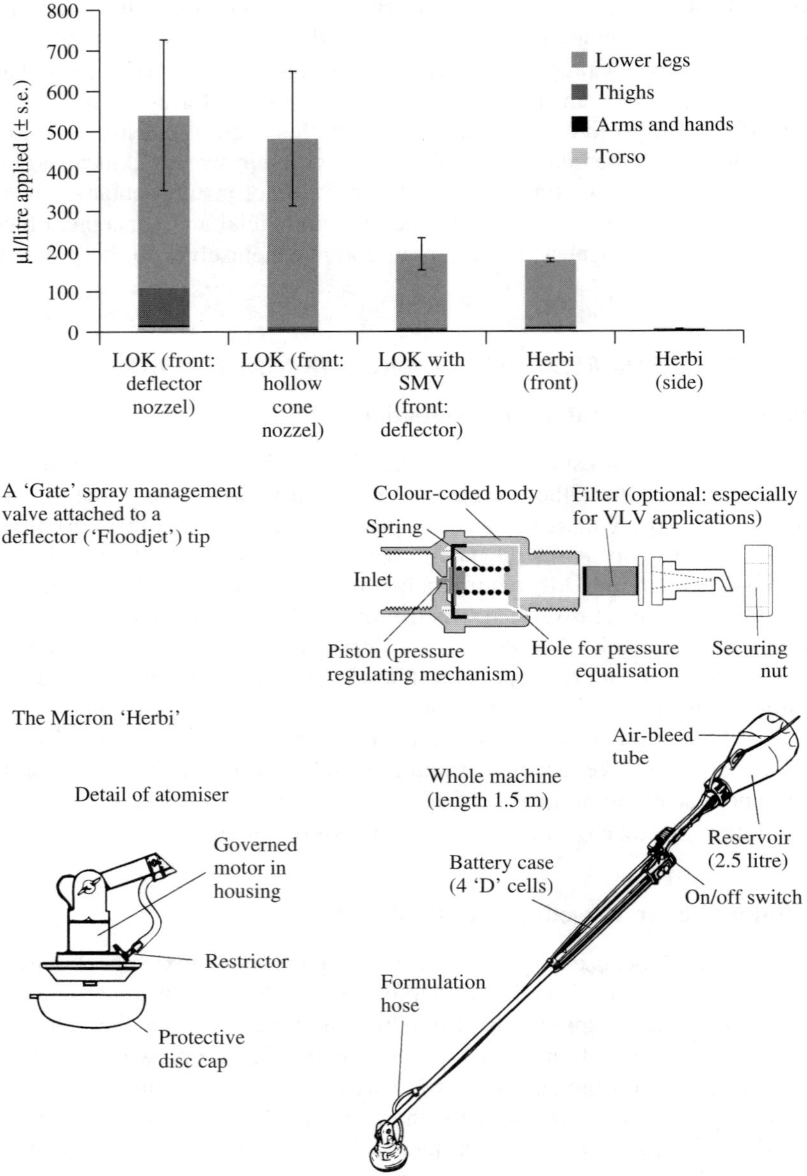

Figure 8.6 Operator contamination with different spraying techniques for herbicide application using lever-operated knapsack (LOK) and 'Herbi' CDA sprayers (data from Thornhill *et al.*, 1995); histogram shows ± s.e. for total deposits. Constant-pressure (spray management) valves (SMV) are now available in a simplified form as illustrated (the 'Gate' valve). Data from Thornhill *et al.* (1995)

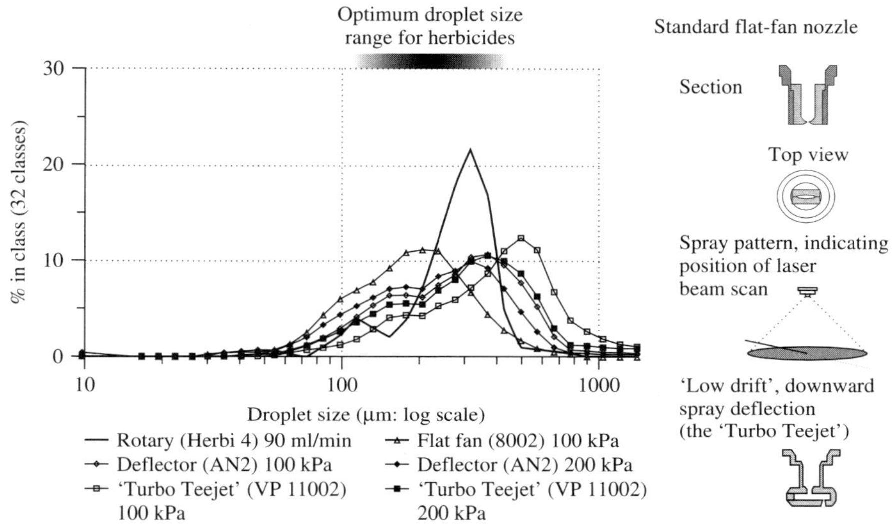

Figure 8.7 Droplet size spectra water +0.1% Agral, atomised by the 'Herbi' rotary atomiser, compared with three hydraulic nozzles. The hydraulic nozzles were scanned diagonally through the centres of the spray fan (as indicated in the centre right diagram)

Multinational companies have shown interest in the development of application technologies, but commitment has not been sustained where intellectual property rights (IPR) in large markets could not be established. Disappointingly few farmers worldwide are aware of alternatives to conventional hydraulic sprayers, which inefficiently use large volumes of water, but remain by far the most important method of pesticide application. Worse still, recent emphasis in application research has focused on the reduction of spray drift (especially in Europe and North America). The most common solution to be implemented to date has been to increase droplet size spectra (without necessarily improving spray quality); thus spray application has probably become generally more inefficient.

Figure 8.7 shows the droplet size spectra produced by nozzles in the study by Thornhill *et al.* (1995). They achieved lowest contamination by controlling pressure at 100 kPa, which, like the newer low-drift nozzles such the 'Turbo Teejet', produce larger spectra than standard flat fan atomisers. However, these settings simply shift the droplet size spectra out of the size range known to be most efficient for pesticides (e.g. Matthews, 1992; Knoche, 1994). The only way to reduce drift and maintain efficient dose transfer is to narrow the droplet spectrum with the optimum range illustrated: using nozzles such as the 'Herbi' rotary atomiser.

Controlled drift spraying is therefore a useful concept, and has become essential for certain types of pest control, following Bals' development of modern, lightweight, spinning disc sprayers, which were first tested in the field in 1968. However, the use of such sprayers has rarely extended beyond the original markets, where a shortage of water led to the development of ultra-low volume (ULV) techniques. CDA represents a very specialised delivery system, with only a limited number of successful cases of implementation. An important development was the modification of spinning disc sprayers for very low volume (VLV) application, where conventional pesticide formulations are mixed with small quantities (5–50 l/ha) of water and thus do away with the need for special, expensive ULV formulations (Nyirenda, 1988). Spraying with larger volumes of water, instead of 0.5–3 l/ha of oil-based formulation, necessitated the development of more robust machines such as the Ulva+ (Figure 8.5; Clayton, 1992). CDA sprayers are the products of medium-scale industrial enterprises, which have been responsible for commercial sustainability over 30 years; they are now well established in certain markets including: smallholder cotton, migrant pest control and herbicide application in amenity areas.

Delivery Systems for Biopesticides: a Quintessential RPU Challenge

A major problem with current pesticide use is the poor standard of application to crops, and techniques used by commercial and semi-commercial developers of biological agents are no exception. Sloppy application of biopesticides may, for instance, be attributed to poor products (that demand the use of large orifice nozzles that emit poor quality sprays) and a belief that putative horizontal transmission mitigates the need for good delivery systems. Laboratory and field tests have clearly shown that, as with chemical pesticides, droplet size and 'coverage' can affect efficacy of agents such as *B. thuringiensis* (e.g. Bryant and Yendol, 1988). With the mycoinsecticide *M. anisopliae* var. *acridum*, it was important to assess spray parameters on potential field performance in preparatory field trials (e.g. Bateman *et al.*, 1998), and to establish that a range of standard application techniques were compatible with the product (e.g. Griffiths and Bateman, 1997).

Unlike many chemical formulations, biopesticides are necessarily suspensions (as opposed to being dissolved in the carrier liquid), and the concept of dose transfer to the target pest in the form of particles must be borne in mind; some of the principles are illustrated in Figure 8.8. At ULV application rates, the quality of both the formulation and emitted spray droplet spectrum are absolutely crucial to maximise the distribution of the very small amounts of biopesticide formulation available per hectare. Figure 8.8a illustrates the droplet size spectra of two atomisers used against locusts and for other migrant pest control. In some countries, exhaust nozzle sprayers (ENS: Sayer, 1959) remain the principal means of applying pesticides to locusts, even though this is difficult to calibrate. The quality of droplets produced by the ENS is inferior to the spectra produced by rotary atomisers such as the X10 (which is fitted to the Micron 'Ulvamast'). Spraying ULV formulations for locust control usually requires a droplet size of

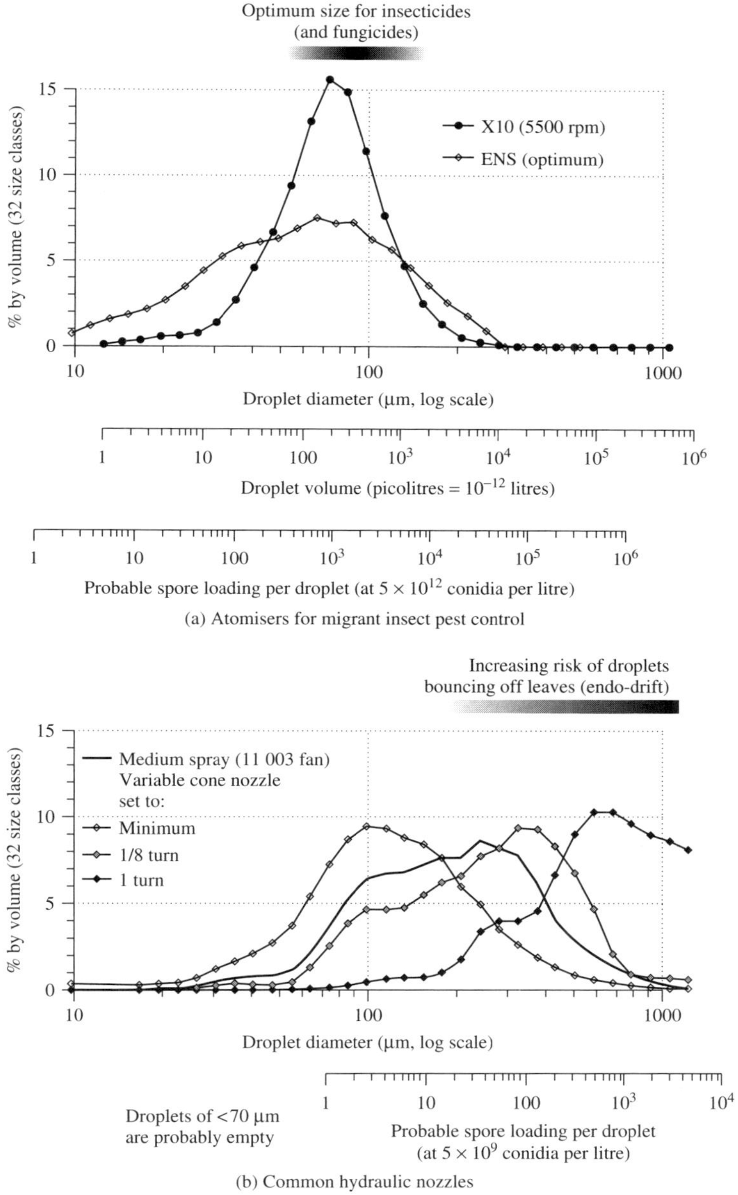

Figure 8.8 Two mycopesticide use scenarios showing droplet size spectra expressed as diameters (in μm) with adjacent volume scale (in pl) and estimated particle distributions in droplet size classes. (a) The atomisers for migrant pest control (usually mounted on vehicles) are both shown applying a blank ULV oil formulation (from) Griffiths and Bateman, 1997); (b) hydraulic atomisers, including different settings of a variable cone nozzle, atomising water +0.1% Agral at 300 kPa (see text)

approximately 40–120 μm diameter (FAO, 1992) and, using these criteria, we can estimate that the CDA technique is at least 1.6 times better, since 80% of the spray volume is within this size band, as opposed to 50% with the ENS (which, unrealistically, was carefully optimised for the purposes of this experiment). Nevertheless, the ENS has a reputation for robustness and, unless very badly adjusted, produces a volume median diameter (VMD) in the optimum, range and does not seriously impair mycoinsecticide viability (Griffiths and Bateman, 1997). With both of these atomisers, the greatest volume of spray consists of droplets in the 50–100 μm range, having volumes of some 65–525 pl and thus expected to contain approximately 330–2600 conidia with a formulation containing 5×10^{12} conidia/litre (typical for 'Green Muscle'). Droplets of >160 μm become increasingly likely to settle near the sprayer and each one "wastes" >10 000 spores, but even with the ENS this only constitutes about 7% of the spray volume.

Unfortunately, CDA is very much the exception rather than the rule. Chapple and Bateman (1997) emphasise the need to develop mycopesticide formulations for conventional application with hydraulic nozzles. Figure 8.8b illustrates some real scenarios in spray application in work that was carried out during an assessment of mycofungicides for cocoa diseases. The bold line represents a medium spray produced by a 'typical' hydraulic nozzle. It is actually a measurement of an 11 003 flat-fan, BCPC reference nozzle measured at 300 kPa, producing a typical wide range of droplet sizes (and also used on tractor-mounted spray booms). For comparison, some measurements are included of a variable cone nozzle: often fitted to manual sprayers in developing countries (and garden sprayers in the UK). When screwed down to its minimum setting it produces a hollow cone spray that is comparable to the fixed cone nozzles shown in Figure 8.4. However, even unscrewing the outer cover slightly, to produce a spray jet (as commonly done when attempting to treat high branches of tree crops) results in a dramatic increase of droplet size. These droplets are highly likely to run off leaves and to be wasted as endo-drift. As before, estimates of spores per droplet have been calculated from droplet volumes and a hypothetical (but observed) tank mixture containing 5×10^{9} spores/litre (i.e. 5×10^{6} spores/ml). By considering spore loading, droplet size spectra and droplet characteristics together, it can be seen that a very high proportion of microbial agents (that are usually costly to prepare) are unlikely to come in contact with their intended biological targets. Increasing the spore concentration will exacerbate this wastage, and decreasing the concentration will increase the proportion of droplets that are likely to be empty. The best approach is to select an appropriate formulation/tank mixture and to apply this using a spraying system that will optimise deposits in the target zone.

A considerable amount of research effort has been expended for the management of spray drift in Europe and North America. However, not only has 'exodrift' been reduced at the expense of 'endodrift' (as discussed above), but conventional drift control may also impose substantial burdens on the development of environmentally benign microbial agents. Since effective delivery systems are crucial for the success of many biopesticides, certain fundamental issues about their future deployment in farming systems should be addressed, including the question: 'should spray drift be an issue at all with biopesticides'?

PRECISION OF APPLICATION IN TIME

MONITORING TECHNIQUES

The need for a high work rate is obvious for large-scale migrant pest control. In field crops, such improvements are only needed when appropriate and timely judgements are made about pest incidence. There is an extensive literature on monitoring techniques and decision making, which has been reviewed by Mumford and Knight (1997) and need not be repeated here. When appropriate decisions are combined with the most suitable (even relatively simple) RPU techniques, yield increases can be spectacular, and perhaps the greatest potential gains will accrue with the development of such combinations for pest control in tropical crops.

In crops such as cotton, overuse of chemical insecticides has resulted in severe resurgence of certain pests, which has hampered cotton growing (e.g. van den Bosch, 1980). On the other hand, the introduction of appropriately applied insecticides in central southern Africa during the 1960s, typically more than doubled yield and improved lint quality (Tunstall and Matthews, 1966). In many countries there is considerable scope for more judicious selection of products and reducing the numbers of sprays (combined with appropriate agronomic practices). The crucial factor is crop scouting; in Africa, simple aids such as the 'peg-board' (Beeden, 1972) have been used effectively over a long period, especially when backed up by an infrastructure for pest scouting, such as the one devised for commercially grown cotton in Zimbabwe (Burgess, 1983). In contrast, Matteson (2000) also stresses the necessity for rice farmers to monitor their fields, but disputes the need for action thresholds, since insecticidal control is often unnecessary and may be counterproductive. The level of pesticide (especially insecticide) use for pest management in many crops remains controversial, but the benefits of monitoring (and better techniques to understand and forecast pest incidence) are not.

Besides simple observation, a number of well-known techniques are available for monitoring pest attacks. Copping (1998) lists 45 pheromones and other attractants, some of which are used as pesticides in their own right, to disrupt mating of lepidopteran and coleopteran pests, or to attract insects towards treated areas or baits. Probably the most frequent use of pheromones is as a monitoring tool to time insecticidal applications. Simple aids such as incandescent light traps, combined with a thorough knowledge of pest biology, also can be most effective. Figure 8.9 shows an example of how incandescent light traps were used for monitoring rice stemborers in Vietnam until the early 1990s. If high catches of

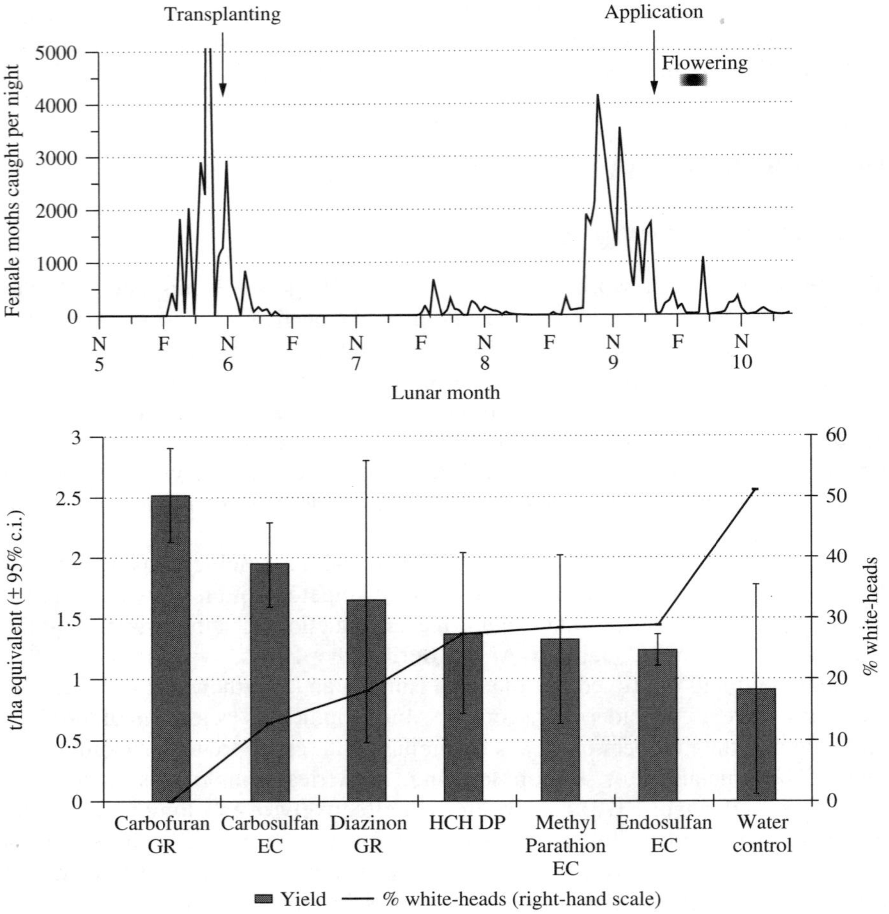

Figure 8.9 A trial on glutinous rice carried out at Hai Hung, Vietnam during the summer crop of 1984 (Bateman *et al.*, 1985). Top: numbers of female *Scirpophaga incertulas* caught in a 100 W incandescent light trap (timing is based on lunar months, where N: new and F: full moon). Bottom: yields (bars) and percentage white-heads (joined lines: right-hand scale), following a single application of various selected insecticides to small plots at the time indicated above

moths (>1000 females) were recorded near to flowering, a single application of insecticide was timed for the emergence of larvae approximately 10 days later. Stewardship was motivated by pest outbreaks and perceived shortages in pesticide supply at the time; however, the technique is rarely used now, with the political changes resulting in a shift from cooperative to family farming. Combined with a single intervention of effective insecticides (and in contrast to older molecules such as HCH and parathion), the yields of highly prized glutinous rice could be more than doubled.

Recent increases in pesticide use in some SE Asian countries have been more dramatic than the global picture illustrated by Figure 8.1, and there is enormous scope there for reducing pesticide use by improving efficiency. This example is admittedly a description of a small plot trial and it is *not* my intention to promote insecticide use on rice. However, it does serve to illustrate the fallibility of precluding all pesticidal control, and the scope for reinvestigating pest management concepts based on combined techniques (i.e. truly integrated pest management).

INCREASING WORK RATE

Among the most important benefits of VLV and ULV spraying techniques are an increase in work rate: the amount of ground (e.g. crop) treated per hour/day (and linked to volume application rate). This enables the *timely, cost-effective* treatment of large areas of land with a minimal amount of effort and formulation. It is an important factor under certain circumstances, for example:

- When the cost of labour is high.
- For a rapid response to scouting: when pest populations exceed action thresholds; this has been shown to be an effective way of reducing sprays on cotton (Matthews, 1996), and a quick response is needed if insect populations are rapidly reproducing.

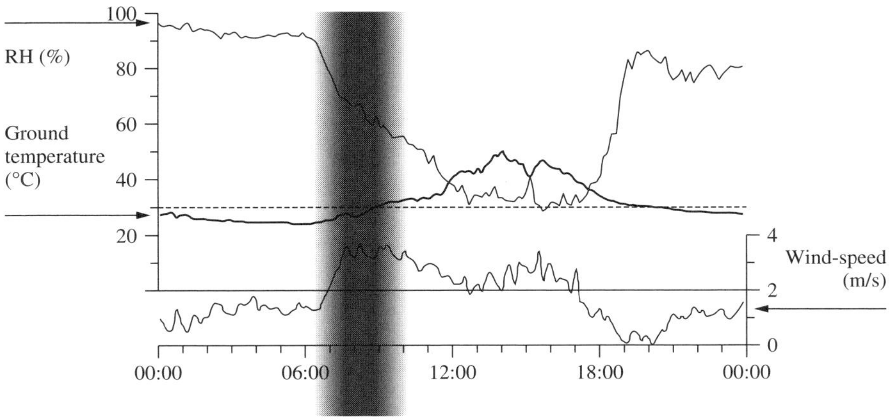

Figure 8.10 The importance of timing in aerial acridid control operations: meteorological conditions over a 24-hour period during the eastern Niger field trial

- Quick response over a wide area is also required in migrant pest control, and high volume application rates with water-based sprays are not feasible for inhospitable terrain.

With migrant pests, a very high work rate is often necessary, since there is only a very limited 'window of opportunity' when spraying can occur. Figure 8.10 shows the climatic conditions in eastern Niger on one day of the aerial application described in Figure 8.3. The effective time available was restricted to an approximately 3 h period in the morning when the wind-speed was >2 m/s and the ground temperature was <30°C. Under these circumstances, ULV drift spraying is the only feasible method of application, and there is considerable pressure to reduce volume application rates well below 1 l/ha in aerial operations. Other techniques for improving work rate, including the use of global positioning systems, are discussed by Dobson (1999).

ECONOMIC AND REGULATORY CONSTRAINTS TO IMPLEMENTATION, AND CONCLUSIONS

Experience with the LUBILOSA Programme has demonstrated what can be achieved with partnerships, involving local, regional and international research organisations, pursuing a concept through to a final product, with commercial collaboration. The Programme has been unusual in several ways – perhaps, most significantly, it represents a sustained research and development investment over a 12-year period. Not only has it brought a practical locust-control option, but also has produced a number of 'spin-offs': for example, contributing to the science of *Metarhizium* taxonomy, biology and field ecology, as well as more applied aspects such as the development of greater understanding of biopesticide delivery systems. Its appeal to donors included: environmentally sound management of pests that have a high political profile, support for marginal communities in sub-Saharan Africa and, most recently, the promotion of links between publicly funded research organisations and small–medium-sized enterprises that manufacture the products.

Biopesticide action contrasts with that of chemicals, but expertise common to both types of pesticide is required for registration, distribution and marketing of products. There are also many common techniques that are useful for optimising delivery systems, but unfortunately with biopesticides this subject is too often a 'no man's land' when it comes to support and implementation (Bateman, 1999). Perhaps most importantly of all, as with chemicals, a consistently high product quality is an essential feature of any reliable biopesticide (e.g. Jenkins and Grzywacz, 2000). Unfortunately, very poor-quality formulations are available in some countries, which may vitiate biopesticide implementation in general. At least some of these problems can be solved by the introduction of 'enabling technologies': crucial techniques or processes that are not necessarily expensive to implement. The International Biopesticide Consortium for Development (IBCD: http://www.biopesticide.org/) has been set up to develop linkages and share information relevant to biopesticide development and commercialisation, including support in: project design, IPR, technology transfer; and identification of new, promising biopesticides.

Other RPU techniques similarly constitute 'orphan' technologies at present, and could become implemented with minimal further research, were they to be understood more widely. Heightened concerns about the environment and food safety present opportunities to develop novel (or rediscover latent) pesticide delivery techniques, some which are being promoted on the http://www.dropdata.net/ web site. Although the characteristics of some of the newer products have improved dramatically in comparison with the early insecticides that gave chemicals such a bad reputation, environmental concerns remain: especially with the more active molecules. Nevertheless, important changes to the nature and range of available of pesticide products may enable the redevelopment of improved application techniques (such as CDA):

- a greatly increased range of a.i.s with a relatively lower mammalian toxicity: enabling their use in a greatly concentrated tank mixture; and similarly
- agents that are effective at much lower rates of application (tens rather than hundreds of g.a.i. per hectare);
- improvements in formulation techniques (especially reductions in solvent levels) and quality control;
- a growing appreciation of closed pesticide transfer systems (e.g. the 'Handy' herbicide containers).

In this Chapter I have tried to show that, having opted for pesticidal interventions, much can be done to reduce their costs and side-effects with better targeting. RPU technologies may be difficult to explain to farmers and require an even greater extension effort than the use of conventional pesticide applications. Amongst the key recurrent problems are: lack of infrastructure (sustainability after development and support for training finishes) and reversion to single technologies ('silver bullets'), rather than truly integrated approaches. Commercial pressures have encouraged the excessive promotion of these 'silver bullets', some of which have proved to be unsustainable practices. To be topical, this may equally apply to the use of a limited number of genes for insect protection in GM crops. RPU requires an alliance between IPM practitioners and providers of technical solutions. All parties need to recognise that:

- In the interests of economic support, quality control and reliability of supply, these technologies are more likely to succeed when provided by commercial concerns rather than rural communities.
- Alliances are needed between research and development organisations and the small–medium-scale industries that have the best track record for sustaining RPU technologies.

- It is essential that farmers and other users understand the biological and technical concepts underlying these technologies.
- A framework of appropriate Government policies is also crucial, but still typified by extremes. Unfortunate examples include: *laissez-faire* regulation in those countries where unprotected smallholders can freely obtain and use toxicity class I products. At the other extreme, there are few biopesticides available in certain European countries – other than *B. thuringiensis* and (unregulated) EPNs. The costs of registering products such as mycoinsecticides have proved prohibitive (even to larger companies) for the niche markets in which they are eminently most suitable.

Pesticide use (rational or otherwise) has been unfashionable with public funding bodies since the early 1990s, with many believing that all pesticide development should be the responsibility of pesticide manufacturers. On the other hand, pesticide companies are unlikely to promote widely better targeting and thus reduced pesticide sales, unless they can benefit by adding value to products in some other way. RPU contrasts dramatically with the promotion of pesticides, and many agrochemical concerns have equally become aware that product stewardship provides better long-term profitability than high-pressure salesmanship of a dwindling number of new 'silver bullet' molecules. RPU may therefore provide an appropriate framework for collaboration between many of the stakeholders in crop protection.

RPU is not a new paradigm; it is unconditionally the subset of IPM practice which recognises that pesticides have been around for more than a generation, and are likely to be around for at least another. Seeing pesticides as dangerous (and conversely biological control as safe) is oversimplistic and sometimes factually wrong. RPU is a combination of old and new ideas to manage real problems such as pesticide resistance, environmental impact, and – of most interest to the farmer – robust but economic pest management. To many perhaps, it will be the potential to make financial savings that will matter most.

REFERENCES

Adams, A.J. and Palmer, A. (1989) Air-assisted electrostatic application of permethrin to glasshouse tomatoes: droplet distribution and its effect upon whiteflies (*Trialeurodes vaporariorum*) in the presence of *Encarsia formosa*. *Crop Protection*, **8**, 40–48.

Adams A.J. Chapple A.C. and Hall F.R. (1990) Droplet spectra for some agricultural fan nozzles, with respect to drift and biological efficiency. In: *Pesticide Formulations and Application Systems: Tenth Symposium. ASTM STP 1078*. (E. Bode, J.L. Hazen and D.G. Chasin (eds), American Society for Testing and Materials, Philadelphia, USA, 156–169.

Bals, E.J. (1969) The principles of and new developments in Ultra Low Volume spraying. *Proceedings of the 5th British Insecticide and Fungicide Conference*, pp. 189–193.

Bateman, R.P. (1999) Delivery systems and protocols for biopesticides. In: *Biopesticides: Use and Delivery*, F.R. Hall and J. Menn (eds), Humana Press, Totowa, NJ, USA; Chapter 24, pp. 509–528.

Bateman, R.P. and Alves, R.T. (2000) Delivery systems for mycoinsecticides using oil-based formulations. *Aspects of Applied Biology* **57**, 163–170.
Bateman, R., Le Truong, Ha Van Chuc, Vu Van Son, Vu Van Minh and Nguyen Duc Long (1985) A field trial to investigate insecticidal control of *Scirpophaga incertulas* with observations on rice crop protection practices in Vietnam. In: R.P. Bateman, *Final Technical Report on Pesticides and Application Methods in the Socialist Republic of Vietnam*. Food and Agriculture Organisation of the United Nations (AGPP internal).
Bateman, R.P., Douro-Kpindu, O.K., Kooyman, C., Lomer, C. and Oambama, Z. (1998) Some observations on the dose transfer of mycoinsecticide sprays to desert locusts. *Crop Protection* **17**, 151–158.
Beeden, P. (1972) The pegboard – an aid to cotton pest scouting. *PANS* **18**, 43–45.
Brent, K.J. and Atkin, R.K. (eds), (1987) *Rational Pesticide Use*. Cambridge University Press, 348 pp.
Bryant J.E., Yendol, W.G. (1988) Evaluation of the influence of droplet size and density of *Bacillus thuringiensis* against gypsy moth larvae (Lepidoptera: Lymantriidae). *Journal of Economic Entomology*, **81**, 130–134.
Bull, D. (1982) *A Growing Problem: Pesticides and the Third World Poor*. Oxfam, Oxford, 192 pp.
Burgess, M.W. (1983) Development of cotton-pest management in Zimbabwe. *Crop Protection*, **2**, 247–250.
Carroll, N.B. (1999) Pesticide development in transition: crop protection and more. In: *Emerging Technologies for Integrated Pest Management*. G.C. Kennedy and T.B. Sutton (eds), American Phytopathological Society Press, St Paul, Minnesota, USA, pp. 339–354.
Carson, R. (1962) *Silent Spring*. Houghton Mifflin, Boston, USA, 368 pp.
Chapple, A.C. and Bateman, R.P. (1997) Application systems for microbial pesticides: necessity not novelty. In: *Microbial Insecticides: Novelty or Necessity?* British Crop Protection Council Proceedings/Monograph Series **68**, 181–190.
Clayton, J.S. (1992) New developments in Controlled Droplet Application (CDA) techniques for small farmers in developing countries – opportunities for formulation and packaging. *Proceedings Brighton Crop Protection Conference – Pests and Diseases*, **1**, 333–340.
Coffee, R.A. (1979) Electrodynamic energy – a new approach to pesticide application. *Proceedings of the 1979 British Crop Protection Conference – Pests and Diseases*, pp. 777–789.
Copping, L.G. (1998) *The Biopesticide Manual*. British Crop Protection Council, Farnham, UK, 333 pp. (2nd Edition due for publication in 2001).
Dobson, H. (1999) Advances in locust spraying technology. *Insect Science and Application*, **19**, 355–368.
Douthwaite, R.J. and Tingle, C.C.D. (1994) DDT in the Tropics: the impact on wildlife in Zimbabwe of ground-spraying for tsetse fly control. *Chatham, UK*: *Natural Resources Institute*, 195 pp.
FAO: Food and Agriculture Organisation of the United Nations (1992) The desert locust guidelines, IV: Control. *Food and Agriculture Organisation Booklet*, P. Symmons (ed.), 64 pp.
FAO: Food and Agriculture Organisation of the United Nations (1997) Evaluation of field trial data on the efficacy and selectivity of insecticides on locusts and grasshoppers. *Report to FAO by the Pesticide Referee Group*, 6th Meeting, Rome, 10–12 December 1996, 21 pp.
Finney, J.R. (1988) World crop protection markets: demisting the crystal ball. *British Crop Protection Conference – Pests and Diseases*, **1**, 3–14.
Georghiou, G.P. (1986) The magnitude of the resistance problem. In: *Pesticide Resistance: Strategies and Tactics*, National Academy Press, Washington DC, USA, pp. 14–43.

Gianessi, L.P. and Carpenter, J.E. (2000). Agricultural biotechnology: benefits of transgenic soyabeans. http://www.ncfap.org/pup/biotech/soy85.pdf: accessed 10/8/2000.
Graham Bryce, I.J. (1989) Environmental impact: putting pesticides into perspective. *Proceedings of the Brighton Crop Protection Conference – Weeds*, 3–20.
Griffiths J. and Bateman, R.P. (1997). Evaluation of the Francome MkII Exhaust Nozzle Sprayer to apply oil-based formulations of *Metarhizium flavoviride* for locust control. *Pesticide Science*, **51**, 176–184.
Gullino, M.L., Leroux, P. and Smith, C.M. (2000) Uses and challenges of novel compounds for plant disease control. *Crop Protection*, **19**, 1–11.
Harman, G.E. (2000) Myths and dogmas of biocontrol: changes in perceptions derived from research on *Trichoderma harzianum* T-22. *Plant Disease*, **84**, 377–393.
Heinrichs, E.A. (1979) Chemical control of brown planthopper. In: *Brown Planthopper: Threat to Rice Production in Asia*, IRRI Publications, Los Banos, Philippines, pp. 145–167.
Himel, C.M. (1974) Analytical methodology in ULV. In: Pesticide application by ULV Methods. *British Crop Protection Council Monograph*, **11**, 112–119.
Hofstein, R. and Chapple, A. (1999) Commercial development of biofungicides. In: *Biopesticides: Use and Delivery* F.R. Hall and J.J. Menn (eds), Humana Press, Totowa, NJ, 77–102.
Jenkins, N.E. and Grzywacz, D. (2000) Quality control of fungal and viral biocontrol agents – assurance of product performance. *Biocontrol Science and Technology*, **10**, 753–777.
Jenkins, N.E. and Vos, J.G.M. (2000) Delivery of Biocontrol technologies to IPM farmers: Vietnam. Co-publication: CAB International, Ascot, UK and UNEP, Nairobi, Kenya; ISBN 92 807 1848 7. 32 pp.
Kayser, H., Kaufmann, L. and Schürmann, F. (1994) Pymetrozine (CGA 215'944): a novel compound for aphid and whitefly control. An overview of its mode of action. *Brighton Crop Protection Conference – Pests and Diseases*, **2**, 737–742.
Knoche, M. (1994) Effect of droplet size and carrier volume on performance of foliage-applied herbicides. *Crop Protection*, **13**, 163–178.
Langewald, J., Ouambama, Z., Mamadou, A., Peveling, R., Stolz, I., Bateman, R., Attignon, S., Blanford, S., Arthurs, S. and Lomer, C. (1999) Comparison of an organophosphate insecticide with a mycoinsecticide for the control of *Oedaleus senegalensis* (Orthoptera: Acrididae) and other Sahelian grasshoppers at an operational scale. *Biocontrol Science and Technology*, **9**, 199–214.
Lisansky, S. (1997) Microbial biopesticides. In: *Microbial Insecticides: Novelty or Necessity?* H.F. Evans, British Crop Protection Council Proceedings/Monograph, **68**, 3–10.
Loevinsohn, M.E. (1987) Insecticide use and increased mortality in rural Central Luzon, Philippines. *The Lancet* (June 1987), 1359–1362.
Lomer, C.J., Bateman, R.P., Dent, D, De Groote H., Douro-Kpindou, O.-K., Kooyman, C., Langewald, J., Ouambama, Z., Peveling, R. and Thomas, M. (1999) Development of strategies for the incorporation of biological pesticides into the integrated management of locusts and grasshoppers. *Agricultural and Forest Entomology*, **1**, 71–88.
Matteson, P.C. (2000) Insect pest management in tropical Asian irrigated rice. *Annual Review of Entomology*, **45**, 549–574.
Matthews, G.A. (1992) *Pesticide Application Methods*. 2nd Edition. Longman, Harlow, Essex, 405 pp.
Matthews, G.A. (1996) The importance of scouting in cotton IPM. *Crop Protection*, **15**, 369–374.
Milaire, H.G. (1995) A propos quelques définitions. *Phytoma – La Défense des Vegétaux*, **475**, 7–9.

Mulqueen, P.J. (1998) Safer formulations of agrochemicals. In: *Chemistry and Technology of Agrochemical Formulations*, D.A. Knowles (ed.), Kluwer Academic Publications, Dordrecht, pp. 121–157.

Mumford, J.D. and Knight, J.D. (1997) Injury, damage and threshold concepts. In: *Methods in Ecological and Agricultural Entomology*, D.R. Dent and M.P. Walton (eds), CAB International, Wallingford, 203–220.

Munthali, D.C. and Wyatt, I.J. (1986) Factors affecting the biological efficiency of small pesticide droplets against *Tetranychus urticae* eggs. *Pesticide Science*, **17**, 155–164.

Newton, I. (1995) The contribution of some recent research on birds to ecological understanding. *Journal of Animal Ecology*, **64**, 675–696.

Nyirenda, G.K.C. (1988) Investigations into very low volume insecticide application to cotton in Malawi. *Crop Protection*, **7**, 153–160.

O'Connell, P.J. and Zoschke, A. (1996) Limitations to the development and commercialisation of mycoherbicides by industry. *2nd International Weed Control Congress, Copenhagen*, 1189–1195.

Peveling, R., Attignon, S., Langewald, J. and Ouambama, Z. (1999) An assessment of the impact of biological and chemical grasshopper control agents on ground-dwelling arthropods in Niger, based on presence/absence sampling. *Crop Protection*, **18**, 323–339.

Pfalzer, H. (1993) Safety aspects and legislation trends. In: *Application Technology for Crop Protection*, G.A. Matthews and E.C. Hislop (ed.), CAB International, Wallingford, pp. 13–33.

Prior, C., Jollands, P. and Le Patourel, G. (1988) Infectivity of oil and water formulations of *Beauveria bassiana* (Deuteromycotina; Hyphomycetes) to the cocoa weevil pest *Pantorhytes plutus* (Coleoptera: Curculionidae). *Journal of Invertebrate Pathology*, **52**, 66–72.

Röling, N.G. and Wagemakers, M.A.E. (1998) *Facilitating Sustainable Agriculture* Cambridge University Press, Cambridge, 318 pp.

Sayer, H.J. (1959) An ultra-low volume spraying technique for the control of the desert locust *Schistocerca gregaria* (Forsk.). *Bulletin of Entomological Research*, **50**, 371–386.

Schnoor, J.L. (ed.) (1992) *Fate of Pesticides and Chemicals in the Environment.* John Wiley & Sons, Ltd., New York, USA.

Thomas, M.B., Wood, S.N. and Lomer, C.J. (1995) Biological control of locusts and grasshoppers using a fungal pathogen: the importance of secondary cycling. *Proceedings of the Royal Society, London, B*, **259**, 265–270.

Thornhill, E.W., Matthews, G.A. and Clayton, J.S. (1995) Potential operator exposure to insecticides: a comparison between knapsack and CDA spinning disc sprayers. *Proceedings of the Brighton Crop Protection Conference: Weeds*, **2**, 507–512.

Tomlin, C.D.S. (ed.) (2000) *The Pesticide Manual.* 12th Edition, British Crop Protection Council, Bracknell, 1250 pp.

Tunstall, J. and Matthews, G.A. (1966) Large scale spraying trials for the control of cotton insect pests in Central Africa. *Empire Cotton Growing Review*, **43**, 121–139.

van den Bosch (1980) *The Pesticide Conspiracy*. Prism Press, Dorchester, UK, 226 pp (originally published in 1978 by Doubleday & Co., USA).

van Emden, H.F. and Peakall, D.B. (1996) *Beyond Silent Spring: Integrated Pest Management and Chemical Safety*. Chapman and Hall, London, 322 pp.

Vos, J.G.M. (ed.) (1998) *Vegetable IPM Exercises: Protocols, Implementation and Background Information.* CABI Bioscience SEARC, 392 pp.

Way, M.J. and van Emden, H.F. (2000) Integrated pest management in practice – pathways towards successful application. *Crop Protection*, **19**, 81–103.

9 Complementary Pest Control Methods

C.H. BELL[1], D.M. ARMITAGE[1] and B.R. CHAMP[2]
Central Science Laboratory, Sand Hutton, York, YO41 1LZ, UK[1]
'Carawah', RMB2, Read Road, Sutton, NSW 2620, Australia[2]

INTRODUCTION

The term 'complementary pest control' can describe two quite distinct spheres of activity. Firstly, it can refer to those activities performed either immediately before or after a pesticide application, or to those performed at regular intervals in between treatments, with the objectives, in both cases, of increasing control efficacy and of reducing the frequency of treatments. Secondly, it can refer to alternative treatments in their own right; in the current context these are of a nonchemical, stand-alone nature.

In practice, the distinction between these two definitions is somewhat blurred, because for many pest or disease problems a combination of measures is taken in the attempt to achieve control, including both chemical and nonchemical components. The term 'integrated pest management' (IPM) is often used to describe such combination treatments, particularly those departing from the use of chemicals. However, it must be emphasised that the components of the IPM system need to be closely defined for each pest-control situation, and thus IPM in itself does not offer a universal panacea for alternatives to chemicals. The term is merely a loose generic descriptor of a wide variety of multicomponent control programmes.

This chapter sets out to give an outline of the complementary pest-control methods that exist, and to provide further details on some in common use in selected situations. Pest control in agriculture and public health is an extensive field, and it is not possible here to present a full account of the multiplicity of methods used. For further detail, the reader should consult one of a number of standard texts on the subject (e.g. Mallis, 1982; Metcalf and Luckmann, 1982; Scopes and Stables, 1989; Pimentel, 1991; Bell *et al.*, 1996).

Complementary measures to a chemical control programme are very wide ranging. For cropping they include soil preparation or substitution, pre-plant treatments (e.g. solarisation, mulching) organic additives, pest detection and diagnosis, preparation of the source crop or seed, selection of resistant varieties, water management, intercropping, biofumigation, crop rotation, as well as post-harvest

Optimising Pesticide Use Edited by M. Wilson
 ISBN: 0-471-49075-X

conditioning and systems approaches in the packing house. For commercial storage and domestic premises, they include environmental control, sanitation, pest detection, various physical additives and physical removal processes, and the use of biological and physical control techniques. Broadly speaking, these adjuncts or alternatives to chemical control can be divided into two categories; those based on living or biologically derived agents, and those based on physical methods or principles.

BIOLOGICAL CONTROL

BIOLOGICAL CONTROL BEFORE CROPPING

The application of beneficial micro-organisms and other agents in pre-plant preparation of soil is a widely practised measure and a mature study area. Recent developments include the establishing of biofumigation systems and the exploitation of antibiotics from bacteria associated with entomopathogenic nematodes. Biofumigation is the amendment of soil with organic matter that releases gases which eliminate or control pests. This may be combined with covering of the soil with plastic or any appropriate system for the purpose of trapping solar energy and raising soil temperatures as well as retaining the gases generated during the process. An example of biofumigation is the incorporation into soil of residues of some brassicas and various Compositae (Bello *et al.*, 1997; MBTOC, 1998). These give off volatile chemicals, such as methyl isothiocyanate and phenethyl isothiocyanate, which have herbicidal, fungicidal, insecticidal and/or nematicidal properties (Gamliel and Stapleton, 1993).

Entomopathogenic nematodes have received considerable attention as biological insecticides because of their ability to search out and kill hosts rapidly (López-Robles *et al.*, 1997; MBTOC, 1998). Recently, the antibiotics produced by their associated bacteria appear promising for control of many plant pathogenic fungi and other soil-borne pathogens (Chen *et al.*, 1994). The bacterium, *Pasteuria penetrans* is effective for control of root-knot nematodes (*Meloidogyne*) in cucumbers and other specific field situations (Tzortzakakis and Gowen, 1994; Stirling *et al.*, 1995).

POST-HARVEST BIOLOGICAL CONTROL

Post-harvest biological control might involve adding to produce, various beneficial fungi, protozoa or bacteria, or parasitoids or predators, in order to control the development of damaging populations of insects or mites. The advantages include the presence of only minimal toxic risks, while the main disadvantages are that biocontrol rarely eliminates pests, may not be able to deal with established infestations and may be rather target-specific. Biocontrol could only have wide efficacy when used in combination. Protozoans, fungi or bacteria may require registration, but native species of parasitoids and predators probably do not.

The predatory mite, *Cheyletus eruditus*, has been applied as a biocontrol agent in empty stores in the Czech republic, used at rates of 2000–3000 per 100 m^2 (Zdarkova and Horak, 1990) or as a top-dressing in ratios of between 1 : 100 and 1 : 1000 predators : prey (Pulpan and Verner, 1965). *Xylocoris flavipes* has also been applied against the storage beetle pests *Tribolium castaneum* (Press *et al.*, 1975) and *Oryzaephilus surinamensis* (Arbogast, 1974), and against the moth *Ephestia cautella* (Press *et al.*, 1982).

Parasitoids applied in the storage environment include *Anisopteromalus calandreae*, where two to three releases at nine-day intervals of the parasitoid at 10× the density of the prey, *Sitophilus oryzae* were recommended (Smith, 1994). *Choetospila elegans* has also been applied against *S. oryzae* (Williams and Floyd, 1971), while *Bracon hebetor* and *Trichogramma pretiosum* (Brower and Press, 1990) have been used against *Ephestia cautella* and *Plodia interpunctella*, and *Ventura canescens* against only the former host.

The protozoans, *Mattesia trogodermae* and *Nosema whitei*, have been used against *Trogoderma glabrum* (Shapas *et al.*, 1977) and *Tribolium castaneum* (Milner, 1972) respectively, while the fungus, *Beauveria bassiana* has been applied against *Sitophilus* spp. and *Tribolium* spp. (Thuy *et al.*, 1995).

The bacterium, *Bacillus thuringiensis*, may have limited application, as resistance could develop in the field after only a few generations (McGaughey, 1994). Nevertheless, variants have been found to have efficacy against *P. interpunctella* (McGaughey, 1985), *Rhyzopertha dominica, T. castaneum, S. oryzae* and *Trogoderma granarium* (Mummigatti *et al.*, 1994).

As many biocontrol agents are host specific, in practice it will be necessary to use many of them in combination. For instance, *Chaoetospila elegans* and *Cephalonomia waterstoni* were employed against a mixed infestation of *Cryptolestes ferrugineus* and *R. dominica* (Flinn *et al.*, 1994). Post-harvest biocontrol is likely to be most useful in dealing with background pest levels in empty stores, as an alternative to fabric treatments or possibly as surface treatments as part of an IPM strategy, based on physical control (cooling and drying).

PHEROMONES

Pheromones are the volatile chemicals that function as messengers for communication between insects. They regulate behaviour, thus providing an opportunity to manipulate the characteristics of pest populations. They are particularly important in reproduction, both in long-range attraction and short-range mate choice (Boake *et al.*, 1996). In the context of pest control in agriculture, stored product protection and domestic situations, pheromones may be used to manage pest abundance, either indirectly as a tool for detection and monitoring of pest populations, or directly in mass trapping to physically remove the insects, in mating-disruption techniques to prevent breeding, or as attracticides to a point where pesticides, pathogens or sterilising agents are used as the control agent.

Pheromones have been reported from a large number of species from six orders of Insecta. There have been many reviews of them in agriculture and plant-protection including those by Jutsum and Gordon (1985), Ridgway *et al.* (1990), Mayer and McLaughlin (1991) and Blum (1996), and in stored product protection by Burkholder and Ma (1985) and Phillips (1994, 1997). These reviews provided lists of materials that have been isolated and identified, as well as those that are used commercially.

There are two basic types of pheromone involved in pest-management systems – sex pheromones and aggregation pheromones. It is necessary also to recognise kairomones, which are interspecific chemical signals that, for example, may be exploited by the parasites (receivers) in locating potential hosts in biological control, and habitat or plant odours that are termed ‘synomones’, as they benefit both their producers (the host plants) and the receivers (parasites of the pests attacking the host plants) by facilitating the location of the pests by the parasites.

SEX PHEROMONES

Sex pheromones are usually emitted by females to attract males for mating. They have been reported from many moths and certain families of beetles, including Anobiidae, Bruchidae and Dermestidae, in which adults are relatively short lived and feed very little or not at all (Burkholder and Ma, 1985). Rarely, the male attracts the female, as in the cabbage looper moth (Landolt and Heath, 1990) and the arctiid moth *Estigmene acrea* (Willis and Birch, 1982).

In Hemiptera, sex pheromones have been identified for the homopteran scale insects; Margarodidae and Diaspididae; the mealy bugs, Pseudococcidae; and Aphididae, and for the heteropteran plant bugs, Miridae; the assassin bugs, Reduviidae; the stink bugs, Pentatomidae; and the shield bugs, Scutelleridae. It is interesting to note that the pheromones are produced by males in taxa characterised by large species, and by females from taxa with relatively small species, suggesting an evolutionary pattern from control pressure by parasitoids (Aldrich, 1996).

Sex pheromone activity may stem from a single major compound, or may involve a mixture of materials of varying activity. Thus, for example, the sex pheromones of the pink boll worm of cotton *Pectinophora gossypiella* are a 1 : 1 mixture of (*Z*, *Z*)- and (*Z*, *E*)-7,11-hexadecadienyl acetates, and those of the oriental fruit moth *Grapholita molesta* a mixture of three components, (*Z*)- and (*E*)-8-dodecenyl acetates and (*Z*)-8-dodecyn-1-ol optimally attractive at the ratio of 95 : 5 of the (*Z*)- and (*E*)-isomers with 3–10% alcohol, while, with the codling moth *Cydia pomonella*, the major component is (*E*, *E*)-8,10-dodecadien-1-ol, but with many additional compounds contributing to the overall effect (Carde and Minks, 1995). Related species may share or at least respond to a common pheromone. Thus, the sex pheromone TDA (*Z*, *E*)-9,12-tetradecadienyl acetate (also known as ZETA), is active not only against *Plodia interpunctella*, but also against at least four other of its pyralid relatives (Brady *et al.*, 1971). Similarly, the anobiids *Stegobium paniceum* and *Anobium punctatum* share stegobinone (Kuwahara *et al.*, 1978), and several *Trogoderma* spp. share (*Z*)-14-methyl-8-hexadecenal (Cross *et al.*, 1976).

AGGREGATION PHEROMONES

Aggregation pheromones are usually produced by males and attract both sexes for mating and to suitable habitats and food sources. The species involved include beetles of the families Bostrichidae, Cucujidae, Curculionidae and Tenebrionidae, which have adults that are relatively long lived and feed substantially (Burkholder and Ma, 1985). As with the sex pheromones, the aggregation pheromones may involve mixtures of materials and related species may share a common pheromone. Thus, the intraspecific dominicalure 1 and dominicalure 2 have been isolated from *Rhyzopertha dominica* (Williams *et al.*, 1981), whereas sitophinone has been isolated from *Sitophilus oryzae* and *S. zeamais* (Schmuff *et al.*, 1984), 4,8-dimethyldecanol from *Tribolium castaneum* and *T. confusum* (Suzuki, 1980), and (*Z*, *Z*)-3,6-dodecadien-11-olide as one of the components from *Oryzaephilus surinamensis* and *O. mercator* (Pierce *et al.*, 1985). Aggregation pheromones have also been reported from mites, Kuwahara *et al.* (1982) identifying lardolure, tetramethyldecyl formate, from *Lardoglyphus konoi*.

PHEROMONES IN PEST MANAGEMENT PRACTICE

Trapping is a valuable tool in pest-management programmes for manipulating pest populations. Pheromones provide a very positive attraction to traps and, because of the extreme sensitivity of insects to these cues, enable infestations to be detected at very low levels, often when visual or other forms of inspection are unsuccessful. The innate specificity of the pheromones enables particular species or species groups to be targeted in trapping programs by using appropriate lures. However, as pheromones are often complex mixtures of related compounds and their stereoisomers can evoke vastly different responses in the species concerned, correct identification, synthesis and blending of the components is essential, as is evaluation of target population responses under field conditions. The pheromone inventories of most important pests have been investigated and, following identification of their critical components, it has been possible to synthesise them in quantities that enable their use in commercial pest control. In stored-product pest control, for example, synthetic pheromones are commercially available for nearly half of the 35 or so species of pests with known pheromones (Phillips, 1994).

Host plants may influence sex pheromone behaviour in insects both in pheromone production and release and in the responses obtained (Landolt and Phillips, 1997). Food odour synergists or other host-plant derived semiochemicals may be incorporated in pheromone trapping systems to improve their performance (Phillips, 1997). The combination of sex pheromone and food odour probably has evolutionary significance in indicating the location of sites offering favourable conditions for reproduction.

The delivery mechanisms for pheromones are crucial. They must be capable of being programmed accurately to achieve the concentration characteristics for the species concerned. They must release the material at appropriate and uniform rates and have capacity consistent with the particular application and operational time frame. Trap design is important, whether the intent is to trap flying insects or both walking and flying insects.

DETECTION AND MONITORING OF PEST POPULATIONS

The most important application of pheromones lies in the detection and monitoring of pest infestations. This information is a critical input for pest-management programmes and associated decision-support systems. A vital issue in these predictive models of pest and infestation behaviour is locating infestations and, to this end, precision targeting of infestations by spatial analysis has proved useful (Brenner, 1997).

Pheromone-based pest monitoring is widely practised in both agriculture and stored-product protection. Some industries use it extensively. Thus, for example, in cotton which has a large pest complex and serious pest problems, many integrated pest-management programmes have been developed, based on monitoring of the boll weevil, *Anthonomus grandis*, the boll worm, *Helicoverpa armigera*, and the pink boll worm, *Pectinophora gossypiella*. Monitoring has also become an integral part of pest control in the food industries, where contamination, not only by whole insects but by fragments of them, is regarded as a public health issue. The sensitivity of pheromone trapping gives timely warning that control measures are necessary, well before other signs are evident. The major pests targeted in the food industries are the pyralid moths. Other extensive applications in stored-product protection are for the tobacco beetle *Lasioderma serricorne* and for *Trogoderma* spp., which are often objects of quarantine, and which, like the pyralid moths, share a common pheromone (Phillips, 1997).

Mass Trapping

Mass trapping is used to reduce pest populations to manageable levels. It is most suited to confined areas such as storage structures, and is most effective at relatively low population densities. Aggregation pheromones are more effective than sex pheromones, because both sexes are attracted to the traps, and not just males. This is particularly evident at high population densities and where there is multiple mating by males. Mass trapping has been trialled with sex pheromones in flour mills to reduce pest populations. Trematerra (1994) used TDA-baited multifunnel traps spaced every 260–280 m^3 and releasing 13 mg TDA daily in order to remove *Ephestia kuehniella*, thus reducing pest numbers to a constant low level. Traps using aggregation pheromones have also been shown to lower bark beetle (Scolytidae) population density in mountain forests, supplementing traditional protection methods (Staryzk, 1996).

Mating Disruption

Disruption of mating can be an effective means of reducing pest populations. It is achieved by flooding the environment with the sex pheromone of the target species, so that mating behaviour is disrupted and reproduction does not occur. It has particular application to free-flying pests such as moths. Carde and Minks (1995) have reviewed the success and constraints of control of moth pests by mating disruption. They discussed the principles of disruption with respect to adaptation, habituation, orientation, competition between artificial and natural releases, and balance of sensory inputs in blended components, and have considered many important species such as the pink boll worm, the oriental fruit moth, and the codling moth, as well as a range of other pests. Sanders (1997) suggested that the disruption is caused by a combination of false-trail following and sensory fatigue, and that high release-rate dispensers are the most effective means of achieving these responses. Ogawa (1997) reviewed the key factors influencing the successful application of the mating disruption technique. These were optimal composition of the blend of disruptant pheromones, as determined by field testing: satisfactory dispenser technology in terms of release method and rates, and dispenser life; air temperature and wind velocity as determinants of release rates and air concentrations of the pheromones; and all of these interacting with the population density of the pest and levels of economic damage thresholds. It was noted that area-wide treatments are most satisfactory, and that they must be applied before emergence of the target species.

As with trapping, the delivery mechanisms for the pheromones are important. They must be able to provide a nominated dose over an extended period, often under adverse weather conditions. Thus, various slow-release methods have been devised for field use and incorporated into the delivery systems. These have involved the use of hollow fibres as reservoirs which can be applied aerially on to the canopy of plants; twisted rope formulations consisting of wire-based sealed polyethylene tubes filled with disruptant that are twisted around the base of plants; other polyethylene or rubber tube or polymeric ampullae dispensers, multilayered laminates of acrylic polymer flakes, resin-treated filter paper; and micro-encapsulated formulations.

Attracticide Systems

Pheromone-based attracticide systems take many forms. Pheromone/pesticide mixtures may be sprayed on to trees for short-term control of pests such as fruitflies, or they may be incorporated into traps or other suitable dispensers to enable continuous control in the field or in storage structures. Trematerra (1994) evaluated 4 cm^2 laminar dispensers baited with 2 mg of TDA and 5 mg of cypermethrin, and spaced every 220–280 m^3 in flour mills – with satisfactory results. Pheromone dispensers may also be exploited to distribute pathogens. Shapas *et al.* (1977) used traps containing pheromones and inoculum of the protozoan *Mattesia trogodermae* to introduce the pathogen into infestations of *Trogoderma glabrum*, while Kellen and Hoffmann (1987) investigated in the laboratory similar methods for dissemination of granulosis virus in populations of *Plodia interpunctella*. Another variant is to use chemosterilants.

Use in Biological Control Programmes

Pheromones play a vital role in biological control. Thus, kairomones improve the host-finding efficiency of parasitoids and predators in biological control programmes. They direct the parasitoids and predators to their hosts or prey, and can be important in both mass-rearing of biological control agents and in release protocols in the field. It is common also for prior experience of exposure to the host kairomone to enhance subsequent response to the host cue (Lewis and Martin, 1990). A classical example of the role of kairomones is the observation that most hemipteran sex pheromones have been exploited by parasites as host-finding kairomones (Aldrich, 1996).

Pheromone-baited traps allow monitoring of pest density throughout parasite release programmes. Thus, it has been demonstrated in field trials that ZETA traps can be used for monitoring phycitine moth populations in biological control programs, using the parasites *Habrobracon hebetor* (Franqui Rivera, 1995), *Trichogramma evanescens* (Scholler and Prozell, 1996) and *Venturia canescens* (Scholler *et al.*, 1997).

INSECT GROWTH REGULATORS

'Insect growth regulators' is a general term for a range of compounds that disrupt normal development in insects, preventing growth of immature stages to reproductive adults. They have a short residual life in soil, water and plants, and an extremely low toxicity to wildlife. Some may affect nontarget insects, however, upsetting often-delicate biological control balances. A wide series of potentially useful insect growth regulators have been evaluated, and a few have found application for managing pest problems affecting humans and livestock. These range from control of mosquitoes and midges, control of face and horn flies in cattle, control of cotton and citrus pests, locusts, defoliating pests in forestry, and household pests such as houseflies, termites, fleas, and cockroaches. The topic has often been reviewed, initially by Staal (1975), and subsequently for particular applications, most recently for citrus pests (Hattingh, 1996), chironomid midges (Ali, 1996), houseflies (Howard and Wall, 1996), cat fleas (Rust and Dryden, 1997) and subterranean termites (Su Nan Yao *et al.*, 1998).

Considerable attention has been focused on direct application to stored commodities, where both residual protection and the quality of the residues involved are of considerable significance. Bengston (1987) discussed this use in detail in relation to stored-product protection, particularly for grain, and, more recently, Oberlander *et al.* (1997) reported on the current status of growth regulators for control of stored-product insects generally. More specifically, there have been many research papers covering different stored-product pests, but attention on practical application appears to have focused primarily on methoprene and grain protection (e.g. Samson *et al.*, 1990).

Juvenile Hormone Analogues

The original concept of insect growth regulators followed the discovery of the juvenile hormone in insects and the realisation that normal growth of insects could be terminally disrupted by introducing mimics (agonists) of juvenile hormones into their metabolism (Williams, 1967). These juvenile hormone analogues were inherently selective, and their activity was confined to insects and other arthropods. Hence, they had negligible acute and chronic mammalian toxicity, a prerequisite for foodstuffs where residual activity was a consideration and could be useful. The most important materials are methoprene, hydroprene, fenoxycarb and pyriproxyfen. They have found application in agriculture, animal health and public health. Of particular significance is the use of methoprene in mosquito control where, due to its selectivity, other forms in the aquatic environment are not harmed.

With foodstuffs, the use of juvenile hormone analogues is of considerable significance where their specificity and low mammalian toxicity are exploited in admixture treatments as grain protectants. Methoprene has been approved by the Joint Meeting on Pesticide Residues of WHO and FAO for application to cereal grains in storage with a maximum residue tolerance of 5 mg/kg. When used for grain treatment, it is applied at rates as low as 1 mg/kg sometimes in admixture with organophosphorus pesticides. Methoprene is a 1 : 1 racemic mixture of the *R*- and *S*-isomers with most of the insecticidal activity coming from the *S*-methoprene. Application rates of 0.6 mg/kg of the *S*-isomer are equivalent toxicologically to rates of 1 mg/kg of the mixture (Bengston *et al.*, 1992). Commercially available formulations contain the racemic mixture. Methoprene has also been widely used as a replacement for phosphine fumigation in protection of stored tobacco against *Lasioderma serricorne* (F.) and *Ephestia elutella* (Hubner) (Manzelli, 1979; Bengston, 1987). Protection for periods up to two years has been reported.

A further application is for control of manure-breeding flies that affect livestock. Garg and Donahue (1989) reviewed the topic of applying methoprene externally to cattle as a spray, or orally in admixture with feed rations, mineral blocks, drinking water, or in sustained-release bolus formulations that are retained in the reticulum and release the methoprene continuously into the faeces. The latter method prevents flies breeding for up to six months. Methoprene may also be admixed with poultry feed to control houseflies in poultry manure.

In domestic pest control, methoprene is used widely to control cat and dog fleas, an application having a residual life of up to nine months (Garg and Donahue, 1989) and has been found effective against house dust mites (Downing *et al.*, 1990). Pyriproxyfen, another promising juvenile hormone mimic, has been extensively studied and found by various authors to have potential for controlling fleas, cockroaches, termites, synanthropic flies, mosquitoes, and chironomid midges. Like methoprene, it has proved particularly useful in control of fleas, giving more than 12 months control of larvae when applied to carpets at 1 mg/m^2 (Kawada and Hirano, 1996). Formulations with other relatively nontoxic pesticides such as the natural and synthetic pyrethroids have improved their efficacy in many domestic applications. Hydroprene has activity somewhat similar to methoprene, but is generally less effective.

Ecdysteroid Mimics

A second group of insect growth regulators, the ecdysteroid mimics, act by interfering with the moulting hormone ecdysone, with consequent prevention of normal metamorphosis. Certain dibenzoyl hydrazines, including tebufenozide, are involved. These materials are essentially experimental, but have the inherent selectivity, low mammalian toxicity and environmental attributes of the juvenile hormone analogues. It has been suggested that they may have potential for use as selective insecticides against Lepidoptera in stored-product protection (Smagghe *et al.*, 1996).

Chitin Synthesis Inhibitors

The third group contains the chitin synthesis inhibitors which also prevent normal moulting of immature stages. They do not mimic insect hormones, and their mode of action is not known. They are principally benzoylphenylureas, and diflubenzuron is the best known. Others are triflumuron, hexaflumuron, flucycloxuron, flufenoxuron, teflubenzuron, and chlorfluazuron. Chitin synthesis inhibitors are effective at comparatively low doses against a wide spectrum of insect pests. It is claimed they have little effect on nontarget organisms or the environment and show promise as a new generation of pesticides (Elek, 1994).

Chitin synthesis inhibitors have proved useful in monitoring – baiting programmes. Thus, with subterranean termites, monitoring stations are used to detect infestations, after which the monitoring stations are replaced with slow-acting baits containing the chitin synthesis inhibitor hexaflumuron to eliminate the termite colonies (Su Nan Yao *et al.*, 1998). Diflubenzuron has been reported to have long residual effectiveness against cat fleas in domestic situations, and triflumuron is particularly effective against cockroaches, provided that there is thorough application to harbourage areas. Whereas control of cockroaches with juvenile hormone mimics is relatively slow, the chitin synthesis inhibitors give much quicker responses, usually at the next moult. The benzoyl urea insect-growth regulators have also found application in barrier treatment of locust swarms (Scherer and Rakotonandrasana, 1993).

Naturally Occurring Growth-Disrupting Compounds

Some natural products used for pest control induce growth-regulatory responses. Nawrot and Harmatha (1994), have discussed these in the context of their better-known antifeedant properties, and Malek and Parveen (1996) have reviewed them for managing *Tribolium* spp. in stored products. Azadirachtin, the active principal in neem (*Azadirachta indica*) disrupts larval and pupal development in *Trogoderma granarium* (Siddig, 1980), and its growth-regulating properties are widely referred to in the literature (e.g. Williams *et al.*, 1996).

Finally, conventional pesticides under certain conditions may induce growth abnormalities which preclude survival of the insects concerned. This can be manifest during pupation and eclosion of adults. It appears to result from sublethal exposure to certain pesticides such as the cyclodienes, where the products of metabolism are toxic and must be excreted to complete the detoxification process. If excretion of the metabolites has not occurred before the pre-pupal stage is reached, they remain in the insect and interfere with morphogenesis.

Some Issues Constraining use of Insect Growth Regulators

As with other biological agents and toxicants, changes in tolerance of insect growth regulators to target species have been reported. These include cross-resistances in nonspecific organophosphorus-resistant *Tribolium castaneum* to the juvenile hormone mimics JH-1, Altozar, Altosid, DMF and Bowers 2B (Dyte, 1972; Dyte *et al.*, 1976). Resistance to juvenile hormone analogues has also been reported in the housefly and tobacco budworm and, in each instance, a strain resistant to a range of traditional pesticides was involved. With the chitin synthesis inhibitors, field resistance has been reported in mosquitoes and houseflies, and at serious levels in *Plutella xylostella* and *Spodoptera littoralis* (Keddis *et al.*, 1986; Lin *et al.*, 1989; Keiding *et al.*, 1991), and laboratory-selected resistances have also been induced in a similar range of pests (El-Guindy *et al.*, 1982; Amin and White, 1984; Ismail and Wright, 1991). In contrast with the juvenile hormone analogues, it has been noted that cross-resistance to other major insecticide groups was minor.

Another constraint in the use of insect growth regulators has been in integrated pest management programmes where economically important bio-control agents may be adversely affected. A classical case has been with citrus in southern Africa, where widespread use of the juvenile hormone analogue pyriproxyfen against organophosphorus-resistant red scale has resulted in outbreaks of pests which were previously unimportant. This resulted in a need for control of these nontarget species with chemicals. The sensitivity of the biocontrol agents to pyriproxyfen has been demonstrated by bioassay to be the cause of the changed status of the pests involved (Hattingh, 1996).

It is common in grain protection for complexes of species to be present or develop during storage, and this must be taken into account in designing systems for their management. In this context, the juvenile hormone analogues have been highly effective against the previously intractable *Rhyzopertha dominica*, as well as the inevitable infestations of the cosmopolitan *Oryzaephilus surinamensis* and *Tribolium castaneum*. Unfortunately, they are less effective against the *Sitophilus* species, possibly due to the development of the immature stages within grains, and do not exercise control at economic treatment rates (Bengston, 1987; Daglish and Samson, 1991; Edwards *et al.*, 1991). Thus, with juvenile hormone analogues, combinations of protectants may be required to ensure control of the full spectrum of pests, as with the methoprene/methacrifos mixtures referred to below. It has been found, however, that the newer juvenile hormone analogue fenoxycarb provides protection against a range of coleopteran pests of grain, including *Sitophilus* spp. (Edwards *et al.*, 1991). With the need to target a range of species, juvenile hormone analogues can be expected to have a role in management systems for grain in storage that is restricted in scope but can be very significant in terms of the scale of the operation.

Thus, Bengston and Strange (1994) report that methoprene has been used successfully as a grain protectant in Australia since 1990, and moreover, that its combination with methacrifos 'has potential to control all currently prevalent strains of grain storage insects infesting cereals present in Australia'. Undoubtedly, then, insect growth regulators have potential to contribute significantly to optimisation of pesticide use in pest management systems in good commercial practice.

Although insect growth regulators are generally accepted as having negligible toxicity to mammals, their residues remain a significant consideration in international trade. They are relatively stable except when exposed to ultra-violet radiation, and problems have resulted where wastes from cotton treated with chlorfluazuron have been used for drought feeding of cattle destined for export as beef to the USA.

COOLING AND FREEZING

The low temperature range affecting arthropods could perhaps be divided into three bands. From the minimum breeding temperature down to the chill coma, insects will be able to feed, but because of their low metabolic rate, be unable to repair accumulating damage. From the chill coma temperature down to the supercooling point, they will be unable to feed and therefore slowly starve. At and below their supercooling temperature, death will occur when the water in their body fluids crystallises (freezes).

Working in the field of grain storage, Evans (1983) identified chill coma temperature ranges of 2.7–5.6°C for the granary weevil *Sitophilus granarius*, 4.4–6.4°C for *C. ferrugineus* and 5.6–10°C for *O. surinamensis*. It is obviously desirable that aeration should reduce grain temperatures to these values to prevent damage by the insects which may live for extended periods above chill coma temperatures.

Smith (1970) found the supercooling point of *C. ferrugineus* to be −17°C without acclimation at an unspecified r.h., while Robinson (1928) found that between 12 and 16% m.c., the supercooling point of *S. granarius* was from −9 to −10°C. Fields (1992) gave the supercooling point of *O. surinamensis* as −16°C. These temperatures are unlikely to be achieved by aeration in the grain stores of cool maritime or warm temperate climates (e.g. UK, France, Spain, Australia, etc.); however, they may be achieved in climates such as those of Scandinavia and Canada.

COLD STORAGE

For many years, cold exposure has been a widely used component of control programmes for perishable commodities such as fresh fruit and vegetables. Exposure times and temperatures are linked to the pest concerned, but need to be chosen after evaluation of effects on the fruit being treated. Usually, the exposure to cold is conducted for a limited period in storerooms or containers in transit, as, for example, the holding of fruit for 10–22 days at −1°C to +2°C to kill tephritid fruit flies on citrus fruit, apples, pears, grapes, stone fruit, carambola, lychees, loquats or kiwi fruit (Gould, 1994). Japan has accepted exposure to cold conditions as a quarantine treatment for four species of fruitflies (Mediterranean, Queensland, Oriental and Melon) on grapes, kiwi fruit and citrus from Australia, Chile, China, Hawaii, Israel, Philippines, South Africa, Spain, and Swaziland, and Taiwan (Kawakami, 1996). Potential quarantine treatments based on cold exposure have also been studied for codling moth *Cydia pomonella* (Moffitt and Burditt, 1989*a*, *b*) and oriental fruit moth *Grapholitha molesta* (Dustan, 1963; Yokoyama and Miller, 1989). Precise records of the temperature and duration of exposure are required to show compliance with phytosanitary treatment specifications for the cold treatment to be accepted as a disinfestation procedure.

AMBIENT COOLING

For durable commodities such as stored grain, cooling using ambient air was originally devised to even out temperature gradients after harvest, in order to minimise moisture movement and associated problems of condensation at the grain surface, with its problems of sprouting and fungal growth (Holman, 1960). Later it was used to lower grain temperatures below the thresholds for insect breeding (Burges and Burrell, 1964), and, finally, it was discovered that if temperatures near zero were to be maintained for a storage season, then infestations could be eliminated (Armitage and Llewellin, 1987).

The theory behind ambient air cooling to prevent pest development is to pass a 'cooling front' through a mass of grain before eggs laid by wandering females on the day that the grain was placed in store, can develop to adulthood. In practice, this means that we have to assume that grain is taken into store at a temperature optimum for insect development (30–35°C) which leaves just 14 days for the grain to be cooled to below 17°C, the threshold for the fastest developing insect, *O. surinamensis*, before it can complete its development (Howe, 1956). It is then necessary to reduce the grain temperature to below 10°C – the threshold for *S. granarius* (Robinson, 1926) and ultimately below 5°C – the threshold for *A. siro* (Cunnington, 1984).

Ambient air cooling is achieved by forcing air through commodities via ducts beneath them. For grain storage, the required rate is about $10\,m^3/h\,t^{-1}$, at which airflow, sufficient air of low temperature can be delivered to achieve the temperature targets outlined during the UK post-harvest months from July onwards (Armitage *et al.*, 1992). The most efficient way to control the hours that fans operate, both to deliver air of optimum temperature and at minimum cost, is to use some form of automatic control, for instance, based on a temperature differential of 4–6°C. However, most grain stores operate on a manual principle, usually switching fans on last thing in the evening and switching them off in the morning. This is a stark contrast with the sophisticated computer-controls utilised in potato stores.

The limitations of cooling by ambient air are that the grain at the surface changes temperature rapidly, permitting insect survival there, and that the surface grain may absorb moisture in damp climates, permitting the development of mite infestations (Armitage *et al.*, 1994).

REFRIGERATION

In many apparently warm climates, such as Australia (Sutherland, 1968) and Israel (Navarro *et al.*, 1969), ambient air cooling may still be appropriate for cooling grain below insects' breeding thresholds, but, in other climates, such as the southern states of the USA, sufficient cool air may not be available immediately after harvest, so that the air may have to be artificially reduced in temperature (Maier *et al.*, 1997).

Grain refrigeration, sometimes referred to as 'chilling' is achieved by passing ambient air over cooling coils. As this raises the r.h., it is usually necessary to heat the air by a few degrees C, to reduce the r.h. again to equilibrium with the grain, *c*.70% r.h. In contrast, potato stores require a high r.h., to prevent 'shrinkage' of the commodity. Refrigerated stores usually require insulation, to prevent rewarming of the product and, because of the extra energy required to cool the air, refrigeration costs are at least three times those of ambient air cooling, although the principles remain the same. As the refrigerator produces waste heat, its operation can be made more cost effective by using the waste heat for drying or to heat a building.

DRYING AND DESICCANTS

Reducing the moisture content of freshly harvested produce may be overlooked as a pest control method, perhaps because the effects of moisture changes on insect biology are underestimated, but often because the engineering aspects of the subject, such as determining the most cost-efficient methods of moisture reduction, often overlook the biological consequences. For cereals and oilseeds, the behaviour in storage of the commodities is often determined by its treatment during drying. The commodity must not only be dried to a moisture content which will prevent growth of fungi and mites, but the speed of drying must be such that the populations of these pests do not exceed acceptable numbers or cause unacceptable quality changes in the commodity.

The relationship between moisture content (the usual market measurement) and relative humidity (the measurement used in most biological studies) is central to the discussion of drying and pests. Any commodity at a given moisture content (m.c.) is in equilibrium with a particular relative humidity (the e.r.h.), and this relationship changes with temperature and depends on whether the grain is picking up (adsorbing) or losing (desorbing) moisture. The method used to measure the m.c. is another obvious source of variation.

To prevent pest development, it is necessary first to determine the lowest r.h. at which the pest can develop, and then determine the m.c. which is in equilibrium with this r.h. For instance, 65% r.h. is sufficient to prevent the development of most fungi and mites, as well as many insects, including psocids and fungus-feeding beetles such as *Ahasverus* and *Typhaea*. At 15°C, this would be about 14.5% m.c. as determined by the ISO oven method. However, if grain at this m.c. were to be shipped to a warmer climate, the e.r.h would rise, even though the m.c. remained the same and fungal, mite and insect activity could recommence. This is particularly relevant in the transport of grain from cool to warm climates.

Control is often thought of merely in terms of immediate death of pests after a treatment, as occurs for instance when an insecticide spray is used against a flying insect. In these terms, lowering the air humidity is unlikely to have such dramatic effects. However, control usually operates in a much subtler fashion, in preventing pests ovipositing, completing their development or even keeping populations below detectable levels. It may be that adults can comfortably survive a given r.h. and even lay eggs, but that most of the eggs die and that none develop into adults.

Relative humidity can affect insect biology in a number of ways, so the following data are of interest:

- time of survival of adults or stages at different humidities;
- oviposition and fertility/productivity;
- time to develop from egg to adult.

The above all combine to affect calculated rates of increase for pest populations.

Desmarchelier (1988) suggested insect rates of increase were linearly related to wet-bulb temperatures, and illustrated this by translating moisture contents (or r.h.) in previously published work into wet-bulb temperatures. By extrapolation, he was then able to define wet-bulb temperatures at which no increase could occur. He found that the wet-bulb temperature permitting zero increase was 16.4°C for *O. surinamensis* based on work by Howe (1956*a*), while for *T. castaneum* it was 17.6°C – based on work by Howe (1956*b*), or 16.3°C – based on work by Shazali and Smith (1986). A wet-bulb temperature of 16°C translates to a dry-bulb temperature of 22.5°C at 50% r.h., to 24.5°C at 40% r.h., to 27°C at 30% r.h. or 30°C at 20% r.h. This suggests that relatively high temperatures are required for insects to breed at low r.h.

This 'wet-bulb hypothesis' is very attractive, and would be most convenient to use as the basis of humidity as a control regime. However, it is to some extent undermined by exceptions. For instance, according to Arbogast (1976) *O. surinamensis* can complete its development at 30°C, 12% r.h., although the gross rate of production (females/female/generation) was only 18% of that at 74% r.h.

The development of the moths, *Plodia interpunctella* and *Ephestia elutella* at 11 humidities was the subject of a study by Imura (1981). Humidity had little effect on eggs laid, % hatch, or pupal mortality. Development of *E. elutella* from egg to adult varied from 37 days at 76% r.h. to 92 days at 33% r.h. *P. interpunctella* took 29 days at 76% r.h. and 71 days at 43% r.h. However, the crucial effect of r.h. was on the death of the developing moths, so that *E. elutella* failed to develop at 23% r.h., due to cumulative mortality of the stages, and *P. interpunctella* failed to complete development at 38% r.h.

The psocid, *Liposcelis bostrychophila* can absorb moisture from the atmosphere at r.h. above about 60%, but below this, death occurs in about 10 days (Knulle and Spadafora, 1969). These authors quote Brown (1951) and Broadhead (1950) as saying that *Liposcelis* will not develop below 60% r.h.

Insects are well adapted to control moisture loss, with their wax-coated, chitinous exoskeleton and the spiracular valves that permit exchange of respiratory gases. However, most pest species of mites, such as *Acarus siro*, are of the order Astigmata, and gas exchange is largely passive, through their thin cuticle, so they are very vulnerable to control by low humidities. Cunnington (1984) has defined the lower humidity for development of *A. siro* as being 60–65% r.h. At 50% r.h. he estimated that the mites would survive for between two days at 30°C to 16–32 days at 10°C.

Humidity may also have an effect on the efficacy of insecticides; for instance, Barson (1983) has shown that chlorpyrifos methyl is more effective at high humidity, especially at lower temperatures.

AMBIENT AIR DRYERS

These depend on the same principles as ambient air cooling, but about ten times the air is required to remove moisture, compared with that required for removing heat. In this case, the principle is to pass a drying front, below the threshold for mite and fungal growth, through the grain, before fungal growth or mite numbers in the upper layers of grain become 'unacceptable'. Mite numbers fall after the passage of the drying front, at a rate that depends on the m.c. achieved, but fungal numbers decline much more slowly (Armitage *et al.*, 1994).

HIGH-TEMPERATURE DRYERS

In these systems, hot air is passed over a relatively thin layer of grain. The maximum grain temperature should not exceed 65°C at 20% m.c., reducing by 1°C for each 1% increase in m.c. to preserve germination and breadmaking qualities. High-temperature dryers can reduce grain m.c. more rapidly than bulk dryers, but the grain has to be efficiently cooled after drying, and this may be problematical at peak harvest time.

INERT DUSTS

Inert dusts have been used traditionally as stored grain protectants, and there is increasing interest in their use as alternatives to chemical control measures. A number of studies demonstrating the efficacy of inert dusts as grain protectants have been reported (Desmarchelier and Dines, 1987), as structural treatments in empty stores (Bridgeman, 1994) and as surface treatments in conjunction with aeration (Nickson *et al.*, 1994). Diatomaceous earths (DEs) have been registered for storage use in USA, Canada, Australia, Japan, Indonesia and Saudi Arabia.

The products are based on inert materials such as silica gel or diatomaceous earth (DE) and contain no insecticides or knockdown agents. They are effective against species resistant to pesticides, and are stable at high and low temperatures (McLaughlin, 1994). In contrast to chemical insecticides, which induce rapid immobilisation and kill, the action of inert dusts is slow, and extended exposure periods of 20 days or more may be required to eliminate an insect population. Most products, at the appropriate concentration, provide protection for at least 12 months (McLaughlin, 1994).

Inert dusts act by physical means, with insect mortality thought to occur by desiccation as a result of the dust adsorbing lipids from the cuticle (Ebeling, 1971). Although insects may accumulate large amounts of dust in their guts during preening, this does not apparently hasten their death (Ebeling *et al.*, 1975).

Several studies have investigated the efficacy of inert dusts against parasitic mites, but knowledge of the effects of inert dusts against stored product mites is very limited. Cook and Armitage (1996) investigated the efficacy of 'Dry-acide', applied to wheat at 1 g/kg and 3 g/kg, against *Acarus siro* and *Glycyphagus destructor*. At 14% moisture content (mc) and 17.5°C, both doses gave complete control of both species. At 16% m.c. and 17.5°C, 3 g/kg was fully effective against *A. siro*, but not against *G. destructor*. Fields and Timilick (1995) found that the product 'Super Insecolo' reduced predaceous and grain-feeding mite species by over 98% when applied to wheat at 50 ppm.

In recent studies on mill treatments in Canada, Fields *et al.* (1997) combined an enhanced DE (Protect-It™) formulation with a regular heat treatment in a commercial plant, resulting in greater mortality of insect bioassays at lower temperatures than those normally targeted, and offering potential savings in energy costs.

Due to the particle size of some dusts, they are considered to be respirable and therefore represent a potential hazard to users (Golob, 1997). Diatomaceous earths may be of marine or freshwater origin, and usually contain about 90% SiO_2 (Golob, 1997). Marine diatoms contain high quantities of crystalline silica, which, if inhaled over a period of time, can result in silicosis and other respiratory diseases. However, dusts of larger particle sizes are relatively safe to use with minimum protective equipment (a simple dust mask).

The use of inert dusts as grain protectants and adjuncts to structural treatments may become increasingly more significant as alternatives to conventional chemicals are sought. Their advantages include effective action against insects and mites, cost effectiveness, high persistence and the leaving of no toxic residues on grain. Disadvantages have included the very high treatment rates required, which affect the physical properties of the grain – particularly bulk density, angle of repose and flow rate (Jackson and Webley, 1994). The newer formulations, however, can be applied with success at much lower dosage rates (Korunic *et al.*, 1996).

HEAT

Heat-treatment technologies are notable as pest-control options that are capable of matching the speed of treatment afforded by fast-acting fumigants. Heating can provide an alternative treatment method to using chemicals, but also can enhance the effectiveness of other treatment methods. For fumigants and controlled atmospheres it does this in three ways: by increasing the diffusion and distribution of gases and hence their powers of penetration, by reducing physical sorption, and by increasing the toxicity or level of stress to target pests. Heat is particularly effective in increasing the efficacy of control using CO_2. In various forms, heat-based control techniques have been widely investigated and are widely used in commercial practice for treatment of perishable and durable commodities, for sterilisation of soil, and for treatment of buildings and structures.

TREATMENT OF FRUIT AND PERISHABLE COMMODITIES BY HOT AIR

The use of heated air seeks to exploit the window between pest elimination and fruit damage, a window which is highly commodity-specific and in some cases too narrow for effective use. Two types of heated air treatments are practised: vapour heat and forced hot air. Vapour heat was the first to be used, and applied hot air saturated with water to the fruit, transferring heat by condensation (Armstrong, 1994). Vapour heat treatments feature a rapid heating phase from ambient, followed by a more gradual increase to the critical end-point temperatures of 43–47°C, depending on commodity and pest sensitivity. Vapour heat is now used commercially for quarantine control of Oriental, melon, Queensland and Mediterranean fruitflies on litchi, papaya, and mango exports from Australia, China, Hawaii, Philippines, Taiwan and Thailand to Japan (Kawakami, 1996).

Forced hot-air treatments have been the subject of active research in parallel with the development of commercial equipment incorporating software capable of giving precise temperature control (Williamson and Winkelman, 1989). The first treatment meeting a quarantine standard was developed to control fruitflies on Hawaiian papayas, and comprised a four-phase heating schedule to bring fruit centres to 47.2°C (Armstrong *et al.*, 1989; Hansen *et al.*, 1990). Heat treatment of apples, pears and cherries using forced hot air for disinfestation of codling moth, combined with a controlled atmosphere and post-harvest cool storage to reduce the duration of the heat treatment, is another candidate for quarantine acceptance (Neven, 1995).

HOT WATER DIPS

Hot water is an excellent medium for heat transfer, and has long been used to reduce pathogens on fruit (Armstrong, 1994). From the 1940s the technique was used in conjunction with dips containing ethylene dibromide. More recently, the technique has been examined for use alone, as a means of fruitfly control. For example, standards have been worked up for mangoes (Sharp and Spalding, 1984), comprising a 65-min dip at 46.1–46.7°C against Caribbean fruitfly, for guava (Gould and Sharp, 1990), comprising 35 min at 46.1°C and for red ginger cut flowers (Chantrachit and Paull, 1998), comprising 12–15 min at 50°C. In many cases a very narrow window exists between pest elimination and product damage, with temperatures around 45°C.

GRAIN DRYERS AND HEAT DISINFESTATION

Commodities need to be dried to the m.c./e.r.h. level unsuitable for fungal or mite growth, and this is often achieved by continuous or hot-air dryers. The temperature selected for drying depends on the initial m.c. to protect against loss of seed and baking quality. For instance, safe temperatures range from about 49°C at 27% m.c. to about 63°C at 15–17% m.c. (Lindberg and Sorenson, 1959).

The time–temperature combinations required to kill insects in grain dryers are often shorter than those that will cause damage to the grain (Ghally and Taylor, 1982), and so disinfestation by heat is a real possibility. Free-living insects in the heated air stream around the grain are likely to experience higher temperatures than the hidden stages inside the grain, and it is the latter that will be harder to kill (Evans, 1987). Australian work using a series of experimental and pilot-scale fluidised or spouted-bed heating systems, indicate that heating to 60°C for a few minutes or to 65°C for a few seconds will kill most species of grain pests (Thorpe and Evans, 1983; Evans *et al.*, 1984; Claflin *et al.*, 1986; Sutherland *et al.*, 1987).

USE OF MICROWAVES

Microwaves heat products from the inside out, selecting regions of higher moisture content. They are thus potentially able to target pests in durable commodities such as grain. Pilot studies have been carried out on grain using rapid heating by microwaves, radiofrequency radiation or infrared radiation (Nelson, 1972; Fleurat-Lessard, 1987; Ingemanson, 1997). Recent tests indicate that selective heating of the infesting insects in stored grain increases nonlinearly at frequencies above 10.6 GHz, and that relaxation processes associated with free water in the insect and increased energy transfer at frequencies of about 24 GHz would produce enhanced selective heating (Halverson *et al.*, 1996; Halverson *et al.*, 1997). The technique is under active investigation.

Frequencies between 25 MHz and 2.45 GHz have been tested against pests of perishable commodities. Seo *et al.* (1970) reduced damage within mangoes by applying repeated 10–15 seconds bursts at 2.45 GHz instead of a continuous treatment, but some mango weevils (*Cryptorhynchus mangiferae* (F.)) were able to survive all treatments that did not damage the fruit. Hayes *et al.* (1984) heated papayas by microwaves (2.45 GHz) until centre temperatures reached 38–45°C, prior to a hot-water dip at 48.7°C. The latter was necessary, because, after the microwave exposure, in other parts of the fruit, temperatures varied by as much as 20°C, so that complete control of oriental fruitfly could not be obtained. Further research is required to see whether uneven heating problems can be solved, and the technique used commercially.

HEAT TREATMENTS OF SOIL – STEAMING AND SOLARISATION

Application of Steam

Steaming is a well established and effective technique for treatment of soil against pathogenic organisms, nematodes, weed seeds and insects, and is extensively used for small-scale field treatments within greenhouses and some small-scale nursery plant operations (Nakano and Botton, 1997). Steaming is also used for small-scale, open-field production of some protected production systems, e.g. bulb and cut flowers, and woody fruit and ornamental plants (Correnti and Triolo, 1998; Rodríguez-Kábana, 1998). The use of steam has yet to be demonstrated to have practical application and economic feasibility where large open areas are to be treated (MBTOC, 1998). Expanded use of steam for large-scale, open-field production systems will also require technological improvements and adequate water supplies.

Early methods for applying steam to soil featured the use of perforated steel pipes, which were buried in the soil to a depth of about 25 cm. Depending on the size of the boiler used to generate steam, an area of 20 to 60 square metres could be treated at one time, after which the pipes were pulled out of the soil and positioned for the next operation (Nederpel, 1979). The technique required a great deal of heavy labour. The introduction of the steam plough – a 3-m-wide device, which applied steam at 40 cm depth in the soil while being pulled through the plot by a powered winch, offered some advance – but problems were still encountered in achieving an even distribution of high temperatures in the treated soil.

A better and still widely practised technique is that of sheet steaming. The soil type and extent of cultivation profoundly influence the efficacy of this technique (Nederpel, 1979). Peat soils are very difficult to sterilise, because of their water-retaining properties. During sheet-steaming operations, a positive pressure is generated which can reduce the seal on the plot. To avoid gas and heat loss, nylon nets have been employed to overlay sheeting. The presence of nets allows a pressure increase of about 10 mm water gauge from the sheet-alone level of 5 mm, which greatly assists heat transfer to deeper soil levels. Improvements in heat retention can be achieved by applying a layer of bubble foil over the sheet (Grossman and Liebman, 1995).

Other measures to improve the penetration of heat to deeper soil levels are the installation of a permanent steam-piping system under the plot (Nederpel, 1979); systems operating under reduced pressure provided by fan action (Runia, 1983); and the use of steam–air mixtures, provided by blowing air into the steam supply (Belker, 1990; Labowsky, 1990). The latter technique permits more control of the temperatures achieved, and is favoured where a soil pasteurisation procedure is desired rather than complete sterilisation.

If the temperatures achieved during steaming are too high, an increased mineralisation of the soil and elimination of all beneficial micro-organisms can occur. Notable effects are the increase in available manganese levels, the enhanced production of ammonia from nitrates – a process which continues after the steaming operation (Sonneveld, 1979), and an increase in inorganic bromide levels (Roorda-van-Eysinga, 1984). With the forthcoming withdrawal of the fumigant methyl bromide, an effective wide-spectrum acting compound, because of concerns over ozone depletion, there is currently much interest in re-establishing steaming facilities for use in glasshouses and other protected crop installations, for treatment of soils and artificial growth media.

Solarisation

Soil solarisation is the trapping of solar radiation under plastic sheeting to elevate the soil temperature of moist soil to levels lethal to soil-borne pests, including pathogens, weeds and arthropods (Katan, 1993). It is most successfully used on heavier soils within arid or semi-arid regions with intense sunshine and minimal rainfall, but, with slightly changed technology, it is effective on sandy soils too (MBTOC, 1998). With appropriate technology, it may also be effective under humid conditions (Chellemi *et al.*, 1997).

HEAT DISINFESTATION OF MILLS

Heating above 50°C has been used to control insects in flour mills for over 100 years. It is used by a number of major food processors as an important part of their pest-control program (Heaps and Black, 1994; Clarke, 1996). There are some important considerations to be taken into account before using heat in structures. For example, some structures cannot tolerate the stresses caused by extreme changes in temperature and differential expansion of structural components, e.g. of concrete, wood and steel. In larger structures, the principal problem is one of achieving a uniform distribution of heat. Insects can sometimes migrate to harbourages in outer walls, deep wall or floor crevices or floor drains, and successfully escape the effect of heat treatment. Some greases may liquefy and must be reapplied after heat treatment. Some products cannot withstand the required temperatures, and may have to be removed and treated separately to prevent the reintroduction of pests. The heat-generating source ideally needs to produce a high-volume airstream at temperatures only marginally above the upper target level. Static heating systems are less effective, because of the need for enhanced circulation to distribute the heat effectively.

IRRADIATION

There has been limited acceptance of pest control using irradiation as an alternative or complementary measure to the use of conventional chemical pesticides. This has been despite the inherent dependability of dosing levels and pest responses, and a freedom from pesticide residues comparable with heat and cold treatments. The current campaign, however, to replace methyl bromide as a fumigant for both general pest control and quarantine applications, as well as the tenfold reduction in its Maximum Residue Level to 50 ppb, have rekindled interest in irradiation as a potential alternative for short-term disinfestation treatments. The Joint UN Food and Agricultural Organisation/International Atomic Energy Agency (FAO/IAEA) Division of Nuclear Techniques for Food and Agriculture and the UN Environment Programme, together with its Methyl Bromide Technical Options Committee, are focusing attention on the global aspects, while in the USA, the Methyl Bromide Alternatives Outreach in cooperation with the US Environmental Protection Agency and the US Department of Agriculture, are sponsoring very extensive interaction of workers in the field through a series of Annual International Research Conferences on Methyl Bromide Alternatives and Emissions Reductions (Anon., 1995, 1996, 1997*a*).

The methodologies for use of irradiation can be broadly classified, according to whether direct irradiation of an infestable commodity is involved, or whether indirect methods of pest management are involved. These methodologies are:

- batch and continuous irradiation of parcels of commodities either individually in their cartons or in pallet loads with automatic or manual loading;

- continuous flow irradiation of bulk commodities;
- sterile male release for pest-population management.

The types of ionising radiation used are the highly penetrating gamma-rays generated in adequately shielded sources containing the radionuclides cobalt-60 and caesium-137; similarly penetrating X-rays generated from machines operated at or below an energy level of 5 MeV, and electron-beam machines operating at or below an energy level of 10 MeV. Of the radionuclides, cobalt-60 is used exclusively, or almost so in commercial irradiators. Caesium-137 is promoted as a method of recycling, but it has not found acceptance by commercial operators. Choice of the irradiation source, gamma or electron beam, will depend on the geometry and penetrability of the material to be treated, the size, capital and operating costs of the source of the radiation, and any environmental constraints on the type of source used. Generally, and within the above criteria, the dose of ionising radiation is the critical factor, not the type of radiation, providing that the penetrating power is adequate. In the context of dose rates, the need for accurate monitoring of dosimetry must be emphasised and data assessed on the accuracy of the monitoring carried out.

Electron beams have very low penetration, so the product does determine the method in some instances. Generally, however, dose rates are not commodity-specific, and are relatively unaffected by temperature change, although there is the problem of attrition, so that there is a maximum–minimum for every dosing situation where a product is treated. In principle, irradiation offers prospect for development of generic doses for control of pests, with a consequent advantage over the commodity-specific heat treatments and the temperature–humidity-sensitive fumigations that are offered as alternatives to current methyl bromide technology for short-term disinfestation. Moreover, as well as treatment times being short, there is no requirement for special post-treatment handling such as tempering or temperature equilibration after heat treatment, or aeration to remove residual gas after fumigation.

DIRECT IRRADIATION – SOME GENERAL CONSIDERATIONS

Direct irradiation of commodities has had a wide range of applications, ranging from microbial sterilisation of disposable medical products, and devitalisation of certain regulated seeds at relatively high doses, to control of sprouting in potatoes, onions, garlic and shallots at lower doses (<100 Gy). In control of pests in infestable commodities, direct treatment also involves comparatively low doses. The criterion for effectiveness is usually prevention of reproduction, particularly in quarantine applications. The US Food and Drug Administration (USFDA) limit for fresh plant food is 1 kGy (the FAO limit is 10 kGy, although this is soon to be unrestricted). FAO working groups have identified 300 Gy as the upper limit, although this may need to be increased slightly for mites.

There has, however, been limited commercial application, and activities generally have been restricted to research and development of technology applicable to foodstuffs, particularly grains, herbs and spices, poultry, and fresh horticultural produce. The most extensive application has been with herbs, spices and condiments, which are widely treated commercially to reduce bacterial counts. The volumes involved are large relative to the quantities traded.

Baraldi (1996) lists the countries in which irradiation is used commercially and the types of product involved. The 25 countries and the various commodities for which irradiation has been approved for disinfestation treatments are listed in the report of the FAO/IAEA Consultants' Meeting in 1997, convened to discuss the 'Role of irradiation as an alternative to methyl bromide fumigation of food and agricultural products' (Anon. 1997*b*). These follow the recommendations of the International Consultative Group on Food Irradiation, established under the aegis of FAO, IAEA and WHO. Table 9.1 summarises current commercial applications of food irradiation. Other commodities are potential subjects for irradiation treatment, e.g. some dried products, including fish (Khatoon and Heather, 1990) and fruit and nuts (Robertson and Baritelle, 1996), but economies of scale and unsuitable sites where treatment is required, preclude or constrain use of the technology. Notwithstanding, the US Animal and Plant Health Inspection Service has recommended a generic dose of 7 kGy as a universal treatment for logs, lumber, and unmanufactured wood products (Griffin, 1996) associated with the lumber trade from Russia.

Although tolerance limits have been available for various foodstuffs, including grains, a major constraint to such use has been regulatory restriction and consumer resistance, based on earlier perceptions that irradiated food was unsafe to eat and was inferior nutritionally. This has been despite WHO declaring in 1987 that irradiation was harmless in doses less than 10 kGy. Recently, however, consumer groups in Europe, as well as the WHO, have accepted irradiated foodstuffs as suitable for human consumption. Indeed, the USFDA has now approved irradiation for use on red meat at dose levels sufficient to control bacteria (Coghlan, 1998), which should provide a valuable catalyst for general acceptance of the technology in other applications, including pest control.

As with genetically modified fruit and vegetables, labelling of irradiated foods has been a prerequisite component of consumer acceptance. This is now feasible and enforceable as the methodology for monitoring the incidence of irradiation in foods is available. Thus, colour-change dosimeters can be integrated into packaging, and tests based on thermoluminescence have been developed for detection of seeds, herbs, spices, fruits, vegetables and shellfish that have been irradiated. Other tests based on electron-spin resonance are applicable to chicken, fish, and some fruit, and tests based on antibody detection are available for the cyclobutanones that result from the breakdown of lipids in irradiated fats. Irrespective of this, and in view of the increasing incidence of outbreaks of food contamination and poisoning, it is now conceivable that the irradiation itself, rather than the labelling, could become the requirement for consumer acceptance. Nevertheless,

Table 9.1 Commercial applications of irradiation

Country	Start date	Products					
		Spices	Dried fruit, vegetables	Fresh fruit, vegetables	Potatoes, onions and garlic	Poultry meat	Other
Argentina	1986	X		Spinach			Cocoa powder
Bangladesh	1993				X		Dried fish
Belgium	1981	X	X				Deep-frozen foods
Brazil	1985	X	X				
Canada	1989	X					
Chile	1983	X	X		X	X	
China	1978	X	X	X	X		Rice, sauces, seasonings, sausages
Croatia	1985	X					Dried beef noodles
Cuba	1987				X		Beans
Czech Republic	1993	X					Dried food ingredients
Denmark	1986	X					
Finland	1986	X					
France	1982	X	X			X	Seasonings, shrimps, frog legs
Germany	1997	X					
Hungary	1982	X			X		Wine corks, enzymes
Indonesia	1988	X					Rice
Iran	1991	X					
Israel	1986	X					Condiments
Italy	1996	X					
Japan	1973				X		
Korea Republic	1986	X					Condiments, garlic powder
Mexico	1988	X					Dry food ingredients
Netherlands	1981	X				X	Egg powder, packaging material
Norway	1982	X					
South Africa	1982	X		X			
Thailand	1989	X			X		Sausages, enzymes
UK	1991	X					
USA	1984	X		X		X	
Yugoslavia	1986	X					

From 'Role of Irradiation as an Alternative to Methyl Bromide Fumigation of Food and Agricultural Products', *Report of FAO/IAEA Consultants Meeting, Vienna, 11–15 August, 1997.*

irradiation remains an emotive issue, based on attitudes to nuclear energy, but this is lessening with time and community education.

The commonly promoted application for direct irradiation in insect control is in export treatment of commodities to meet phytosanitary restrictions on movement of insects in international trade (IAEA, 1992). This movement is usually regulated by inspection of the commodity at the time of export, usually with further inspection on receipt in the importing country. A significant disadvantage of irradiation for such export disinfestation protocols is that usually not all of the insects are killed immediately by treatment at commercial dose levels. Data on the times required to achieve complete mortality of 15 stored product species have been collated from the literature by Banks and Fields (1995), indicating periods from 7 to >70 days in adult coleopterans dosed at 300 and 500 Gy. The generally accepted requirement for grain is 200–250 Gy. A holding period after treatment is thus necessary if the commodity is to be declared free of living insects by inspection at the time of export. This restricts the free flow of commodities for rapid outloading, and the detention may increase the opportunities for reinfestation.

Alternatively, if the treatment was carried out at the time of export, its effectiveness would have to be assumed, and could not be verified by inspection before export. This is not in the spirit or intent of the International Plant Protection Convention (Champ and Winks, 1982), and verification of treatment for quarantine or contractual arrangements would have to be done by documentation and/or colour dosimetry. For irradiation, this will require a change in approach, and there are now moves to base infestation assessment on 'prevention of reproduction'. In this context, the US Animal and Plant Health Inspection Service, in association with the North American Plant Protection Organization, has drafted a Regional Phytosanitary Standard for Use of Irradiation as a phytosanitary treatment (Griffin, 1996).

IRRADIATION OF GRAIN

Irradiation of grain has been extensively researched over the past 40 years. Effects on baking quality have been an integral part of these studies (e.g. Wooten, 1985). The early studies in England and Australia of Cornwell and Bull (1960) and Cornwell (1966) of the United Kingdom Atomic Energy Authority, and subsequently in the USA by Brower, Tilton and their colleagues (Brower and Tilton, 1985; Tilton and Brower, 1987) at the US Department of Agriculture Stored Products, Insects Research and Development Laboratory at Savannah, Georgia, provided a sound basis for establishing dose/response data, but did not lead to significant adoption of the technology in the grain industries for either microbial or general pest control, simply because irradiation was logistically more difficult and economically unattractive when compared with fumigation.

The studies usually have found that treatment of grain is only economical if carried out at central facilities in the handling networks, where capital costs can be written off against the high throughput rates and full utilisation of the facility. Indeed, the required economies may only be achieved in some circumstances by treatment of grain at both inloading and outloading of the storage, with export or regional subterminals offering the most appropriate opportunities for use of the technology. In separate studies, the USSR developed electron-beam technology for commercial application to continuous-flow systems for grain (Zakladnoi *et al.*, 1982), and installed facilities domestically and advertised units for sale in the technical literature. Other applications have been in the treatment of small quantities of rice, beans, and other pulses in China, Indonesia and Thailand for market trials that had positive consumer acceptance. Indeed, recently, van Graver (1997) reported that irradiation was used commercially under government supervision in Indonesia to disinfest between 2000–8000 tonnes of rice per month.

Overall, irradiation has not been adopted widely for pest control in grain. It is of interest that, subsequently and somewhat paradoxically, many of the reports of research on this sophisticated and capital-intensive technology appear to originate from the least developed countries. This has resulted in some measure from misdirected development assistance programs.

The principal target species in grain treatment are various coleopterans which are either free-living or develop within individual grains. Other pests such as lepidopterans and mites can be somewhat more tolerant of irradiation but to minimise treatment rates in practice dose levels are based on the more important beetle species. The object of treatment is to prevent reproduction and/or feeding, rather than achieve short-term (immediate) kills. Reproduction can be inhibited immediately in beetles at 200–250 Gy, and is associated with damage to the midgut epithelium and digestive function, which prevents feeding and leads eventually to death from starvation (Brower and Tilton, 1985; Ignatowicz, 1996*a*, and many other papers). This is a common response to exposure to ionising radiation in a wide range of species, including field pests. As well as sterility and accelerated mortality, the sublethal effects concern inhibition of development and alteration of mating, flight and general locomotor behaviour (Tilton and Brower, 1983). It should be noted also that Lepidoptera may require higher doses to achieve the same inhibition of reproduction.

As indicated earlier, the delay in killing insects is a significant disadvantage in treatment of grain for export, particularly in continuous-flow applications, and rearrangement of loading schedules may be necessary if freedom from living insects at time of loading is a contractual or regulatory requirement. Indeed, if such freedom from insects is a constraint, irradiation techniques seem more appropriate for subterminal storage situations where throughputs can be sufficiently high and holding periods can be achieved without interrupting normal operations. Moreover, a further constraint to continuous-flow applications has been the flow rates required in commercial practice. These may be as high as 4000 tonnes per hour, and would certainly require capital-intensive investment for a facility designed to handle throughputs at these levels. Nevertheless, as a physical treatment, irradiation has the inherent advantages of reliability if used in a properly designed facility with adequate treatment protocols and of freedom from residues.

A related application for direct irradiation is for treatment of jute bags infested with *Tribolium confusum*. Small batches containing up to 40 bags can be treated with doses of 2 kGy as a replacement for methyl bromide or phosphine fumigation (Ignatowicz, 1996*b*). Such treatments are not necessarily restricted to used bags from grain or grain products, and can be used for similar purposes in a wide range of applications.

IRRADIATION OF FRESH HORTICULTURAL PRODUCE

Methyl bromide fumigation is currently a widely used treatment for quarantine and general pest control in horticultural produce moving in trade. The relatively short shelflife of the often highly perishable commodities require alternative control measures to involve short treatment times, while retaining the consistency in response associated with methyl bromide. Treatment of fresh horticultural produce by irradiation has been researched widely, particularly for quarantine disinfestation (Heather, 1993). Doses up to 1 kGy have been approved by the US Food and Drug Administration (USFDA, 1986) and have been permitted for such use since 1989 by the USDA's Animal and Plant Health Inspection Service (USDA, 1989). It thus offers immediate prospects for application, including as a replacement for methyl bromide fumigation (Moy, 1988). Such treatment has been shown to be highly efficacious against many pests, including fruitflies, codling moth and light brown apple moth. Regulations are now in place permitting use of radiation disinfestation for Hawaiian fruits being shipped into mainland USA, and a commercial facility is planned to facilitate irradiation of agricultural products (Borsa *et al.*, 1997).

The fruitflies, including the Mediterranean fruitfly, and various Australian species, are of particular concern in both quarantine and commercial contexts, and can be controlled by doses as low as 75–100 Gy (Heather *et al.*, 1991; Burditt, 1992). These doses do not result in adverse reactions in the fruit (Moy, 1985; Heather, 1993). Based on very extensive research, generic doses of 150 Gy have been recommended for fruitfly in all fruit by the Task Force on Irradiation, as a Quarantine Treatment convened by The International Consultative Group on Food Irradiation. Similarly, a generic dose of 300 Gy has been recommended for all pests other than fruitfly, based on data for 13 pests from six orders of Insecta, and a criterion of inhibition of reproduction.

With the economies of scale, treatment of bulk commodities would only be possible in central packing houses, as with potatoes and onions and possibly citrus and pome fruits. Packaging, however, does not constrain the penetrating nature of irradiation, and treatment of batches of cartoned product may be either in batches or continuous-flow systems. In-transit times may be very short if fruit and vegetables are air-freighted, and delays in reaching end-point mortality will again assume importance, particularly in quarantine situations. Quarantine treatments normally require a nil tolerance at inspection, plus a security treatment for critical pests. Under these circumstances, irradiation is a practical alternative to methyl bromide, heat and cold if it is combined with a quality control programme to ensure that the pest is below the level of detection at packing. A similar problem exists with cut flowers, where air-freighting is the normal mode of transport. It would be essential for prior arrangements to be made to cope with this problem.

It seems apparent that treatment costs associated with irradiation will be higher than for methyl bromide fumigation. Estimates of cost increases range from 2 to 3.8 times the cost of treatments with methyl bromide (Aegerter and Folwell, 1996). Nevertheless, with economies of scale, the technology can have practical application (Forsythe and Evangelou, 1994). To achieve the full potential for use of irradiation in control of post-harvest insect problems in fresh horticultural produce, more research and development is required to confirm dosing protocols and in development of handling methodology.

STERILE INSECT RELEASE TECHNIQUES

Following the realisation that irradiation induced sterilisation of insects, one of the earliest developments was investigations on control of pest populations by release of sterile males in numbers sufficient to control natural increase in the population, and either achieve its eventual extinction or suppress pest numbers to economic levels (Bushland and Hopkins, 1951, 1953; Lindquist, 1955; Knipling, 1955; Lindquist and Knipling, 1957). The demonstration of eradication of the screw worm (*Callitroga hominivorax*) from the island of Curacao in the Netherlands Antilles by weekly release of males sterilised by gamma rays (Baumhover *et al.*, 1955) was spectacular proof of the potential of the method. Subsequently, the so-called sterile insect release technique (SIRT) has been researched for a wide range of field and storage pests, including blowflies, fruitflies, pink bollworm, and flour beetles and moths (Cornwell, 1966; Hooper, 1970; Tilton and Brower, 1987; Henneberry and Clayton, 1988). This technology, however, is peripheral to the discussion on complementary pest-control methods, and is not dealt with in detail here.

FUTURE PROSPECTS FOR USE OF IRRADIATION

For many years, and particularly in its earlier developmental stages, irradiation was a technology looking for applications and finding a community reluctance to embrace and exploit its immense potential. This is now changing rapidly. Properly used, the method gives reliable responses, which are a prerequisite for quarantine applications. Currently, there do not appear to be any economic incentives as such for irradiation, and other factors will determine whether it becomes used as a treatment method – factors such as unavailability of any other treatment for whatever reason. The USDA–APHIS recently permitted test shipments of

Hawaiian produce of atemoya, bananas, litchi, melon, orange, papaya, rambutan and starfruit to the Chicago region, for irradiation to control pests. Shipments have increased each year, starting with six shipments in 1995, 16 in 1996, and 37 in 1997 (Wong, 1998).

MODIFIED ATMOSPHERES

Atmospheres based on carbon dioxide (CO_2) or nitrogen (low oxygen) offer alternative treatments to fumigation for insect and mite pest control in a wide range of commodities. They are ineffective against fungal pests. They are unlikely to be used for disinfestation purposes where a fast turnaround is necessary, unless combined with other factors such as high pressure or raised temperature. Whereas CO_2 atmospheres have to be provided from bulk supplies, either from cylinders or cryogenic tanks depending on the scale of the operation, N_2-based atmospheres can also be generated on site.

Currently, the most economical method for the generation of a controlled atmosphere is by combustion of natural gas or propane, where the product atmosphere comprises 87–89% N_2 and 10–12% CO_2. Other methods involve the separation of N_2 from air. One such system is based on a principle for gas separation, known as pressure-swing adsorption or PSA. In this process, N_2 is derived from compressed air passed through two beds of molecular-sieve coke. The nitrogen and oxygen are separated due to their different rates of adsorption, with the nitrogen passing through the bed and into a holding tank. Another system is based on filtration of air through membranes which differentiate between O_2 and N_2. The semi-permeable membranes are mounted in separators containing thousands of hollow-membrane fibres and capable of withstanding high pressures. Incoming air at 100–150 psi (7–10 bar) passes along the fibres, and O_2 permeating through the fibre walls is removed from the vessel by suction from the free space in the separator vessel.

PEST CONTROL USES ON PERISHABLES

For perishable commodities, the use of modified atmospheres to extend post-harvest life and quality of fruit and vegetables has been practised for over 60 years, and the technique is now widely used in combination with lowered temperature and raised humidity. The atmospheres used for storage (up to 10% CO_2, $< 5\%$ oxygen) include those known to be effective against insect pests, but the low-temperature combination acts against the efficacy of the treatment. For fungistatic action, the addition of 5–10% carbon monoxide to the atmosphere provides protection for commodities that cannot tolerate high CO_2 levels (Kader and Ke, 1994).

Work on the effect of controlled atmospheres (CA) on different insect pests of perishable commodities was recently reviewed by Carpenter and Potter (1994).

These authors also carried out the first commercial CA quarantine treatment in the export of asparagus from the US to Japan, featuring a 4.5-day exposure to 60% CO_2, followed by transport at 0–1°C. On a much smaller scale, atmospheres immediately surrounding the commodity can be modified to kill pests, using polyfilms or coatings made from wax or cellulose-based compounds (Hallman *et al.*, 1994). Shrink wrapping has the effect of enabling an atmosphere to develop that is lethal to pests.

CA USE ON DURABLE COMMODITIES

Treatments with controlled or modified atmospheres based on CO_2 and nitrogen N_2 offer an alternative to fumigation with toxic gases for insect pest control in all durable commodities, but usually at an increased cost as compared with conventional chemical treatments.

Currently, there is limited use on cereals (Nataredja and Hodges, 1990), an increasing use for the treatment of museum artefacts (Newton *et al.*, 1996), and use of CO_2 on a number of high-value commodities under high pressure (Prozell *et al.*, 1997).

Structures for use with controlled or modified atmospheres must be well sealed to achieve effective levels and keep gas usage and expense to acceptable levels (Mann *et al.*, 1997). The use of a continuous flow of controlled atmosphere, such as that provided by combustion of propane or from a nitrogen generator, can allow somewhat less gastight enclosures to be treated (Bell *et al.*, 1993, 1997). These low-oxygen atmospheres need to reduce oxygen levels below 1% for effective action. Carbon dioxide atmospheres do not rely on achieving anoxic conditions to be effective, and are usually applied to achieve a minimum of about 60% CO_2 (in air) in the treatment enclosure. The efficacy of CAs can be improved for most developmental stages by lowering the r.h., because of a reduced ability to restrict desiccation during the exposure (Navarro and Calderon, 1974).

The use of low-oxygen or high-CO_2 atmospheres requires exposure times to be extended from two to three weeks at 25°C to six to eight weeks at 15°C, if tolerant species such as the weevils *Sitophilus* spp. are present, with the high-CO_2 atmospheres offering the slightly shorter exposure times (Conyers and Bell, 1997). Such timescales may be acceptable for grain in storage and for museum artefacts, but can create problems for other uses. The exposure times required for high-CO_2 atmospheres can be reduced if the treatment is combined with raised temperature or alterations of pressure. Complete kill of a range of stored-product pests can be obtained within 48 h by raising the temperature to 38°C (Jay, 1986).

Exposure under vacuum reduces exposure times for complete kill of the lesser grain borer *Rhyzopertha dominica* at about 20°C from several weeks to a few days (Jay, 1986; Locatelli and Daolio, 1993), while exposure at raised pressures of 10 and 15 bar achieved complete kill of all stages of the granary weevil *Sitophilus granarius* (L.) within 16 h and 8 h respectively at this temperature (Le Torc'h and Fleurat Lessard, 1991). At 20 bar and 20°C, control of a variety of stored-product pests can be achieved within 8 h (Prozell *et al.*, 1997). Some facilities operating at 25 bar are in commercial use to treat cocoa, herbs and spices (MBTOC, 1998).

HERMETIC STORAGE

Hermetic storage is employed routinely for dealing with the world's largest crop, in the production of silage. For storage of cereal grains, it offers the prospect of long-term preservation of quality, provided that humidity levels are low (Pixton *et al.*, 1975). In commodities of a moisture content in equilibrium with an r.h. above 70%, fungal growth will occur, and, if the container in which this occurs is sealed, the oxygen atmosphere will be used up and replaced with carbon dioxide, presumably due to the activities of aerobic storage fungi. At 1–2% oxygen, 15–40% carbon dioxide, the yeasts *Hansenula* and *Candida* take over (Clarke and Hill, 1981). After this, there may be an anaerobic fermentation associated with lactic acid bacteria (>0.91 a.w.) and yeasts (Nichols and Leaver, 1966). The grain now takes on a characteristic taint, so that it becomes unfit for breadmaking and the germ dies. The CO_2 can rise to nearly 95% of the intergranular atmosphere anaerobically (Hyde and Oxley, 1960) but, in practice, there is loss of CO_2 and entry of O_2 due to leaks and pressure changes within a silo, so the CO_2 normally stabilises at 15–25% (Hyde and Burrell, 1969). Carbon dioxide may also be heavily absorbed by the grain (Mitsuda *et al.*, 1973).

CONCLUSIONS

Complementary pest control is an integral part of pest management, and it is assuming greater importance as pressures mount on the continuing use of chemicals. The two drivers for reduction of chemical use are concerns over the environment, including global warming, ozone depletion, acid rain, groundwater contamination, and long-term persistence in soil, and concerns over human and animal health, namely the problems of applicator safety, residues in foods, spray drift and allergic or immunological responses. There is thus increasing emphasis on the need to justify the use of each chemical pest-control agent as a vital component of a pest-management system, rather than as an independent combative action. Hence, complementary pest-control methods are becoming increasingly important in the pest-control scene, backed by decision-support systems and an increasing awareness of pest problems and control measures among staff in the broad fields of food, agriculture and health.

REFERENCES

Aegerter, A.F. and Folwell, R.J. (1996) Economic alternatives to the use of methyl bromide in the postharvest treatment of selected fruits. In: *1996 Annual International*

Research Conference on Methyl Bromide Alternatives and Emissions Reductions, 4–6 November 1996, Orlando, Florida, USA, p. II.
Aldrich, J.R. (1996) Sex pheromones in Homoptera and Heteroptera. In: Studies on Hemipteran Phylogeny, Schaefer, C.S. (ed.) Entomological Society of America, Lanham, MD, USA, pp. 199–233.
Ali, A. (1996) A concise review of chironomid midges (Diptera: Chironomidae) as pests and their management. *Journal of Vector Ecology* **21**, 105–121.
Amin, A.M. and White, G.B. (1984) Resistance potential of *Culex quinquefasciatus* against insect growth regulators methoprene and diflubenzuron. *Entomologia Experimentalis et Applicata* **36**, 67–76.
Anon. (1995) *1995 Annual International Research Conference on Methyl Bromide Alternatives and Emissions Reductions, 6–8 November 1995, San Diego, California, USA*, MBAO (Methyl Bromide Alternatives Outreach), Fresno, California.
Anon. (1996) *1996 Annual International Research Conference on Methyl Bromide Alternatives and Emissions Reductions, 4–6 November 1996, Orlando, Florida, USA*, MBAO (Methyl Bromide Alternatives Outreach), Fresno, California.
Anon. (1997*a*) *1997 Annual International Research Conference on Methyl Bromide Alternatives and Emissions Reductions, 3–5 November 1997, San Diego, California, USA*, MBAO (Methyl Bromide Alternatives Outreach), Fresno, California.
Anon. (1997*b*) Role of irradiation as an alternative to methyl bromide fumigation of food and agricultural products. *Report of FAO/IAEA Consultants' Meeting, Vienna 11–15 August 1997*, FAO/IAFA, Vienna.
Arbogast, R.T. (1974) Suppression of *Oryzaephilus surinamensis* on shelled corn by the predator, *Xylocoris flavipes*. *Journal of the Georgia Entomological Society* **11**, 63–67.
Arbogast, R.T. (1976) Population parameters for *Oryzaephilus surinamensis* and *O. mercator*. Effect of relative humidity. *Environmental Entomology* **5**, 738–742.
Armitage, D.M. and Llewellin, B.E. (1987) The survival of *Oryzaephilus surinamensis* (L.) (Coleoptera: Silvanidae) and *Sitophilus granarius* (L.) Coleoptera: Curculionidae) in aerated bins of wheat during British winters. *Bulletin of Entomological Research* **77**, 457–466.
Armitage, D.M., Wilkin, D.R. and Cogan, P.M. (1992) The cost and effectiveness of aeration in the British climate. *Proceedings of the 5th International Working Conference on Stored-Product Protection* **3**, pp. 1925–1933.
Armitage, D.M. Cogan, P.M. and Wilkin, D.R. (1994) Integrated pest management in stored grain: combining surface insecticide treatments with aeration. *Journal of Stored Products Research* **30**, 303–319.
Armstrong, J.W. (1994) Heat and cold treatments. In: *Insect Pests and Fresh Horticultural Products: Treatments and Responses*. Paull, R.E. and Armstrong, J.W., (eds), CAB International, Wallingford, UK, pp. 103–119.
Armstrong, J.W., Hansen, J.D., Hu, B.K.S. and Brown, S.A. (1989) High temperature, forced-air quarantine treatment for papayas infested with tephritid fruit flies. *Journal of Economic Entomology* **82**, 1667–1674.
Banks, J. and Fields, P. (1995) Physical methods for insect control in stored-grain ecosystems. In: *Stored Grain Ecosystems*, Jayas, D.S., White, N.D.G., and Muir, W.E., (eds) Marcel Dekker, New York.
Baraldi, D. (1996) Irradiation in the preservation of spices, herbs and condiments. *Informatore Fitopatologico* **46**, 23–27.
Barson, G. (1983) The effect of temperature and humidity on the toxicity of three organo-phosphorus insecticides to adult *Oryzaephilus surinamensis*. *Pesticide Science* **14**, 145–152.
Baumhover, A.H., Graham, A.J., Bitter, B.A., Hopkins, D.A., New, W.D., Dudley, F.H. and Bushland, R.C. (1955) Screw-worm control through release of sterilized flies. *Journal of Economic Entomology* **48**, 462–466.

Belker, N. (1990) Environmentally friendly soil sterilization using steam. *Deutscher-Gartenbau* **44**, 672–677.

Bell, C.H., Price, N.R. and Chakrabarti, B. (1996) *The Methyl Bromide Issue*, John Wiley & Sons, Ltd, Chichester, England.

Bell, C.H., Chakrabarti, B., Conyers, S.T., Wontner-Smith, T.J. and Llewellin, B.E. (1993) Flow rates of controlled atmospheres required for maintenance of gas levels in bolted metal farm bins. In: *Proceedings, International Conference on Controlled Atmosphere and Fumigation in Grain Storages, Winnipeg, June 1992*, Navarro, S. and Donahaye, E. pp. 315–325, Caspit Press, Jerusalem.

Bell, C.H., Conyers, S.T. and Llewellin, B.E. (1997) The use of on-site generated atmospheres to treat grain in bins or floor stores. *Proceedings of the International Conference on Controlled Atmosphere and Fumigation in Stored Products, Nicosia, Cyprus, April 1996*, Donahaye, E.J., Navarro, S. and Varnava, A., pp. 263–271, Printco, Cyprus.

Bello, A., González, J.A. and Tello, J.C. (1997) La biofumigación como alternativa a la desinfección del suelo. *Horticultura Internacional* **17**, 41–43.

Bengston, M. (1987) Insect growth regulators. In: *Proceedings of the 4th International Working Conference on Stored Product Protection, Tel Aviv, Israel, September 1986*, Donahaye, E. and Navarro S. (eds), pp. 35–46.

Bengston, M., Koch, K. and Strange, A.C. (1992) Development of insect growth regulators in Australia and South-east Asia. In: *Proceedings of the 5th International Working Conference on Stored Product Protection, Bordeaux, France, September 1992*, Fleurat-Lessard, F. and Ducom, P. pp. 485–490.

Bengston, M. and Strange, A.C. (1994) Recent developments in grain protectants for use in Australia. In: *Stored Product Protection. Proceedings of the 6th International Working Conference of Stored Product Protection, Canberra Australia, April 1994*, Highley, E., Wright, E.J., Banks, H.J. and Champ, B.R. (eds) pp. 751–754, CAB International, Wallingford, UK.

Blum, M.S. (1996) Semiochemical parsimony in the Arthropoda. *Annual Review of Entomology*, **41**, 353–374.

Boake, C.R.B., Shelly, T.E. and Kaneshiro, K.Y. (1996) Sexual selection in relation to pest-management strategies. *Annual Review of Entomology* **41**, 211–229.

Borsa, J., Kotler, J., Kunstadt, P., Reid, B. and Fraser, F. (1997) Irradiation for quarantine disinfestation: practical aspects and options. In: *1997 Annual International Research Conference on Methyl Bromide Alternatives and Emissions Reductions, 3–5 November 1997*, San Diego, USA, p. 67.1.

Brady, U.E., Tumlinson, J.H. III, Brownlee, R.G. and Silverstein, R.M. (1971) Sex pheromone of the almond moth and the Indian meal moth: cis-9, trans-12-tetradecadienyl acetate. *Science* **171**, 801–804.

Brenner, R.J. (1997) Spatial analysis in precision targeting for integrated pest management: concepts and processes. In: *1997 Annual International Research Conference on Methyl Bromide Alternatives and Emissions Reductions, 3–5 November 1997*, San Diego, California, USA, pp. 50.1–50.4.

Bridgeman, B.W. (1994) Structural treatment with amorphous silica slurry: an integral component of GRAINCO's IPM strategy. *Proceedings of the 6th International Working Conference Stored-Product Protection* **2**, 628–630.

Broadhead, E. (1950) A revision of the genus *Liposcelis* Motschulsky with notes on the position of this genus in the order Corrodentia and on the variability of ten *Liposcelis* species. *Transactions of the Royal Entomological Society London* **101**, 335–388.

Brower, J.H. and Press, J.W. (1990) Interaction of *Bracon hebetor* and *Trichogramma pretiosum* in suppressing stored-product moth populations in small inshell peanut storages. *Journal of Economic Entomology* **83**, 1096–1101.

Brower, J.H. and Tilton, E.W. (1985) The potential of irradiation as a quarantine treatment for insects infesting stored food commodities. In: *Radiation Disinfestation of Food and Agricultural Products*, Moy, J.H. (ed.) 75–86. University of Hawaii at Manoa, Honolulu.

Brown, C.D. (1951) Temperature and Relative Humidity as Factors in the Biology of *Liposcelis dionatoria* (Muller) (Corrodentia: Atraopidae). Ph.D. Thesis, Iowa State College, 72 pp.

Burditt, A.K. (1992) Effectiveness of irradiation as quarantine treatment against various fruit fly species. In: *Use of Irradiation as a Quarantine Treatment of Food and Agricultural Commodities*. International Atomic Energy Agency, Vienna.

Burges, H.D. and Burrell, N.J. (1964) Cooling bulk grain in the British climate to control storage insects and to improve keeping quality. *Journal of the Science of Food and Agriculture* **15**, 32–50.

Burkholder, W.E. and Ma, M. (1985) Pheromones for monitoring and control of stored-product insects. *Annual Review of Entomology* **30**, 257–272.

Bushland, R.C. and Hopkins, D.E. (1951) Experiments with screw-worm flies sterilized by X-rays. *Journal of Economic Entomology* **44**, 725–731.

Bushland, R.C. and Hopkins, D.E. (1953) Sterilization of screw-worm flies with X-rays and gamma-rays. *Journal of Economic Entomology* **46**, p. 648.

Carde, R.T. and Minks, A.K. (1995) Control of moth pests by mating disruption: successes and constraints. *Annual Review of Entomology* **40**, 559–586.

Carpenter, A. and Potter, M.A. (1994) Controlled atmospheres. In: Sharp, J.L. and Hallman, G.J. (eds), *Quarantine Treatments for Pests of Food Plants*. Westview Press, Boulder, Colorade, pp. 171–198.

Champ, B.R. and Winks, R.G. (1982) Infestation and degradation – the grain drain. In: *Grain – Trade, Transportation and Handling. Proceedings of the 1st International Grain, Trade, Transportation and Handling Conference, 15–16 June, 1982, London*. C.S. Publications Ltd, Worcester Park, England.

Chantrachit, T. and Paull, R.E. (1998) Effect of hot water on red ginger (*Alpinia purpurata*) vase life. *Postharvest Biology and Technology* **14**, 77–86.

Chellemi, D.O., Rich, J.R., Barber, S., McSorley, R. and Olson, S.M. (1997) Adaption [*sic.*] of soil solarisation to the integrated management of soilborne pests on tomato under humid conditions. *Phytopathology* **87**, 250–258.

Chen, G., Dunphy, G.B. and Webster, J.M. (1994) Antifungal activity of two *Xenorhabdus* species and *Photorhabdus luminescens,* bacteria associated with the nematodes *Steinernema* species and *Heterorhabditis megidis*. *Biological Control* **4**, 157–162.

Claflin, J.K., Evans, D.E., Fane, A.G. and Hill, R.J. (1986) The thermal disinfestation of wheat in a spouted bed. *Journal of Stored Products Research*, **22**, 153–161.

Clarke, J.H. and Hill, S.T. (1981) Microflora of moist barley during sealed storage in farm and laboratory silos. *Transactions of the British Mycological Society* **77**, 557–565.

Clarke, L.W. (1996) Heat treatment for insect control. *Proceedings of the Workshop on Alternatives to Methyl Bromide, Toronto, Canada, May 30–31*, pp. 59–65.

Coghlan, A. (1998) Out of the frying pan. *New Scientist* **2115**, 14–15.

Conyers, S.T. and Bell, C.H. (1997) The effect of modified atmospheres on the juvenile stages of six grain beetles. *Proceedings of the International Conference on Controlled Atmosphere and Fumigation in Stored Products, Nicosia, Cyprus, April 1996*, Donahaye, E.J., Navarro, S. and Varnava, A., pp. 73–81.

Cook, D.A. and Armitage, D.M. (1996). The efficacy of an inert dust on the mites *Glycyphagus destructor* Schrank and *Acarus siro* L. *International Pest Control*, **38** (6), 197–199.

Cornwell, P.B. (ed.) (1966) *The Entomology of Radiation Disinfestation of Grain*. Pergamon Press, New York, 236 pp.

Cornwell, P.B. and Bull, J.O. (1960) Insect control by gamma irradiation: an appraisal of the potentialities and problems involved. *Journal of the Science of Food and Agriculture*, **11**, 754–758.

Correnti, A. and Triolo, L. (1998) Cases of strawberry production without MB in Italy. In: *Alternatives to Methyl Bromide for the Southern European Countries*. Bello, A., González, J.A., Arias, M. and Rodríguez-Kábana, R. (eds), DG XI, EU, CSIC, Madrid, pp. 151–158.

Cross, J.H., Byler, R.C., Cassidy, R.F. Jr, Silverstein, R.M., Greenblatt, R.E., Burkholder, W.E., Levinson, A.R. and Levinson, H.Z. (1976) Porapak-Q collection of pheromone components and isolation of (Z)- and (E)-14-methyl-8-hexadecenal, potent sex attracting components, from the frass of four species of *Trogoderma* (Coleoptera: Dermestidae). *Journal of Chemical Ecology* **2**, 457–468.

Cunnington, A.M. (1984) Resistance of the grain mite, *Acarus siro* L. (Acarina, Acaridae) to unfavourable conditions beyond the limits of its development. *Agriculture, Ecosystems and Environment* **11**, 319–339.

Daglish, G.J. and Samson, P.R. (1991) Insect growth regulators as protectants against some insect pests of cereals and legumes. In: *Proceedings of the 5th International Working Conference on Stored Product Protection, Bordeaux, France, September 1992*, Fleurat-Lessard, F. and Ducom, P. (eds) pp. 509–514.

Desmarchelier, J.M. (1988) The relationship between wet-bulb temperature and the intrinsic rate of increase of eight species of stored-product Coleoptera. *Journal of Stored Products Research* **24**, 107–113.

Desmarchelier, J.M. and Dines, J.C. (1987) Dryacide treatment of stored wheat: its efficacy against insects and after processing. *Australian Journal of Experimental Agriculture* **27**, 309–312.

Downing, A.S., Wright, C.G. and Farrier, M.H. (1990) Effects of five insect growth regulators on laboratory populations of the North American house-dust mite *Dermatophagoides farinae*. *Experimental and Applied Acarology* **9**, 123–130.

Dustan, G.G. (1963) The effect of standard cold storage and controlled atmosphere storage on survival of larvae of the oriental fruit moth. *Journal of Economic Entomology* **56**, 167–169.

Dyte, C.E. (1972) Resistance to synthetic juvenile hormone in a strain of the flour beetle, *Tribolium castaneum*. *Nature* **238**(48) p. 49.

Dyte, C.E., Forster, R. and Aggarwal, S. (1976) The rust-red flour beetle. Resistance to juvenile hormone mimics. *Pest Infestation Control*, **1971–73**, 79–80.

Ebeling, W. (1971) Sorptive dusts for pest control. *Annual Review of Entomology*, **16**: 123–158.

Ebeling, W., Reierson, D.A., Pence, R.J. and Viray, M.S. (1975) Silica aerogel and boric acid against cockroaches: external and internal action. *Pesticide Biochemistry and Physiology* **5**, 81–89.

Edwards, J.P., Short, J.E. and Abraham, L. (1991) Large-scale evaluation of the insect juvenile hormone analogue fenoxycarb as a long-term protectant of stored wheat. *Journal of Stored Products Research* **27**, 31–39.

Elek, J.A. (1994) *The response of Rhyzopertha dominica and Sitophilus oryzae to chitin synthesis inhibitors in stored grain*. Ph.D. Thesis, Australian National University, Canberra.

El-Guindy, M.A., El-Rafai A.R.M. and Abdel-Sattar, M.M. (1982) The pattern of cross-resistance to insecticides and juvenile hormone analogues in a diflubenzuron-resistant strain of the cotton leaf worm *Spodoptera littoralis* Boisd. *Pesticide Science* **14**, 235–245.

Evans, D.E. (1983) The influence of relative humidity and thermal acclimation on the survival of adult grain beetles in cooled grain. *Journal of Stored Products Research* **19**, 173–180.

Evans, D.E. (1987) Some biological and physical restraints to the use of heat and cold for disinfesting and preserving stored products. *Proceedings of the 4th International Working Conference on Stored Product Protection*, 149–164.

Evans, D.E., Thorpe, G.R. and Sutherland, J.W. (1984) Large scale evaluation of fluid bed heating as a means of disinfesting grain. *Proceedings of the 3rd International Working Conference on Stored Product Protection* 523–530.

Fields, P.G. (1992) The control of stored-product insects and mites with extreme temperatures. *Journal of Stored Products Research* **28**, 89–118.

Fields, P. and Timilick, B. (1995). *Efficacy Assessment of Super Insecolo*. Report for Hedley Pacific Ventures Ltd., Vancouver, Canada.

Fields, P.G., Dowdy, A. and Marcotte, M. (1997) *Structural Pest Control: the Use of an Enhanced Diatomaceous Earth Product Combined with Heat Treatment for the Control of Insect Pests in Food Processing Facilities*. Report prepared for Environment Bureau, Agriculture and Agri-Food Canada and United States Department of Agriculture, 25 pp.

Fleurat-Lessard, F. (1986) Utilisation d'un attractif de synthèse pour la surveillance et le piegeage des pyrales Phycitinae dans les locaux de stockage et de conditionnement des denrees alimentaires végétales. *Agronomie*, **6**, 567–573.

Flinn, P.W., Hagstrum, D.W. and McGaughey, W.H. (1994) Suppression of insects in stored wheat by augmentation with parasitoid wasps. *Proceedings of the 6th International Working Conference on Stored-Product Protection*, **2**, 1103–1105.

Forsythe, K.W., Jr. and Evangelou, P. (1994) *Costs and benefits of irradiation versus methyl bromide fumigation for disinfestation of U.S. fruit and vegetable imports*. Staff Report – Economic Research Service, United States Department of Agriculture, No. 9412, ii + 39 pp.

Franqui–Rivera, R.A. (1995) Behaviour, patterns of seasonal activity, and cold tolerance in *Bracon hebetor* Say (Hymenoptera: Braconidae). Ph.D. Thesis, University of Wisconsin–Madison, 139 pp.

Gamliel, A. and Stapleton, J.J. (1993) Characterization of antifungal volatile compounds evolved from solarized soil amended with cabbage residues. *Phytopathology* **83**, 99–105.

Garg, R.C. and Donahue, W.A. (1989) Pharmacologic profile of methoprene, an insect growth regulator, in cattle, dogs, and cats. *Journal of the American Veterinary Medicine Association* **194**, 410–412.

Ghally, T.F. and Taylor, P.D. (1982) Quality effects of heat treatment of two wheat varieties. *Journal of Agricultural Engineering Research* **27**, 227–234.

Golob, P. (1997) Current status and future perspectives for inert dusts for control of stored product insects. *Journal of Stored Product Research* **33**, 69–79.

Gould, W.P. (1994) Cold storage. In: *Quarantine Treatments for Pests of Food Plants*, Sharp, J.L. and Hallman, G.J. (eds), Westview Press, Boulder, Colorado, pp. 119–132.

Gould, W.P. and Sharp, J.L. (1990) Hot water immersion treatment for guavas infested with Caribbean fruitfly (Diptera: Tephritidae). *Journal of Economic Entomology* **85**, 1235–1239.

Graver, J.E.S. van (1997) Methyl bromide – are there alternatives for grain disinfestation. In: *Proceedings of the 18th ASEAN Seminar on Grains Postharvest Technology, Manila, 11–13 March 1997*.

Griffin, R.L. (1996) Update: regulatory policies for the use of irradiation as a phytosanitary measure. In: *1996 Annual International Research Conference on Methyl Bromide Alternatives and Emissions Reductions*, 4–6 November 1996, Orlando, Florida, USA, p. 69.

Grossman, J. and Liebman, J. (1995) Alternatives to methyl bromide – steam and solarization in nursery crops. *The IPM Practitioner* **XVII(7)**, 1–12.

Hallman, G.J., Nisperos-Carriedo, M.O., Baldwin, E.A. and Campbell, C.A. (1994) Mortality of Caribbean fruit fly (Diptera: Tephritidae) immatures in coated fruits. *Journal of Economic Entomology* **87**, 752–757.

Halverson, S.L., Plarre, R., Burkholder, W.E., Bigelow, T.S. and Misenheimer, M.E. (1996) SHF and EHF microwave radiation as a pesticide alternative for stored

products. *Annual International Research Conference on Methyl Bromide Alternatives and Emissions Reductions. November 4–6, Orlando, Florida, 1996*, pp. 55-1–55-4.
Halverson, S.L., Plarre, R., Bigelow, T.S. and Lieber, K. (1997) Advances in the use of EHF energy as a fumigant for stored products. *Annual International Research Conference on Methyl Bromide Alternatives and Emissions Reductions. November 3–5, 1997. San Diego, California*. pp. 111-1–111-4.
Hansen, J.D., Armstrong, J.W., Hu, B.K.S. and Brown, S.A. (1990) Thermal death of oriental fruit fly (Diptera: Tephritidae) third instars in developing quarantine treatments for papayas. *Journal of Economic Entomology* **83**, 160–167.
Hattingh, V. (1996) The use of insect growth regulators in integrated pest management of citrus in Southern Africa. *Citrus Journal* **6**, 14–17.
Hayes, C.F., Chingon, H.T.G., Nitta, F.A. and Wang, W.J. (1984) Temperature control as an alternative to ethylene dibromide fumigation for the control of fruit flies (Diptera: Tephritidae) in papaya. *Journal of Economic Entomology* **77**, 683–686.
Heaps, J.W. and Black, T. (1994) Using portable rented electric heaters to generate heat and control stored product insects. *Association of Operative Millers, Bulletin*, **July 1994**, pp. 6408–6411.
Heather, N.W. (1993) Irradiation as a quarantine treatment of agricultural commodities against arthropod pests. In: *Management of Insect Pests: Nuclear and Related Molecular and Genetic Techniques*, pp. 627–639, International Atomic Energy Agency, Vienna.
Heather, N.W., Corcoran, R.J. and Banos, C. (1991) Disinfestation of mangoes with gamma irradiation against two Australian fruit flies (Diptera: Tephritidae). *Journal of Economic Entomology* **84**, 1304–1307.
Henneberry, T.J. and Clayton, T.E. (1988) Effects of gamma radiation on pink bollworm (Lepidoptera: Gelechiidae) pupae: adult emergence, reproduction, mating, and longevity of emerged adults and their F1 progeny. *Journal of Economic Entomology* **81**, 322–326.
Holman, L.E. (1960) Aeration of grain in commercial storages. *Marketing Research Report* **178**. Washington: US Department of Marketing Services Research Division.
Hooper, G.H.S. (1970) Sterilization of the Mediterranean fruitfly. In: *Sterile-Male Technique for Control of Fruit Flies*, 3–11, International Atomic Energy Authority, Vienna.
Howard, J. and Wall, R. (1996) Control of the house fly, *Musca domestica*, in poultry units: current techniques and future prospects. *Agricultural Zoology Reviews* **7**, 247–265.
Howe, R.W. (1956*a*) The biology of the two common species of *Oryzaephilus* (Coleoptera; Cucujidae) *Annals of Applied Biology* **44**, 356–368.
Howe, R.W. (1956*b*) Effect of temperature and humidity on the rate of development and mortality of *Tribolium castaneum* (Herbst) (Coleoptera: Tenebrionidae) *Annals of Applied Biology*, **44**, 341–355.
Hyde, M.B. and Burrell, N.J. (1969) Control of infestation in stored grain by airtight storage or by cooling. Proceedings of the 5th Brighton Insecticides and Fungicides Conference 412–419.
Hyde, M.B. and Oxley, T.A. (1960) Experiments on the airtight storage of damp grain. I. Introduction; effect on the grain and the intergranular atmosphere. *Annals of Applied Biology* **48**, 687–710.
IAEA (1992) *Use of Irradiation as a Quarantine Treatment of Food and Agricultural Commodities*. International Atomic Energy Agency, Vienna.
Ignatowicz, S. (1996*a*) Starvation as a cause of death in storage beetles following gamma irradiation. *Roczniki Nauk Rolniczych, Seria E*, **T. 25, Z. 1/2**, 105–111.
Ignatowicz, S. (1996b) Efficacy of electron beams in radiation disinfestation of jute bags. *Roczniki Nauk Rolniczych, Seria E*, **T. 25, Z. 1/2**, 125–131.
Imura, O. (1981) Effect of relative humidity on the development and oviposition of four phycitid moth pests associated with stored-products. *Report of the National Food Research Institute* **38**, 106–114.

Ingemanson, M.O. (1997) MIDS™ Infrared technology – effective, benign, affordable. *Annual International Research Conference on Methyl Bromide Alternatives and Emissions, November 3–5, 1997, San Diego, California*, **113**, p. 1.

Ismail, F. and Wright, D.J. (1991) Cross-resistance between acylurea insect growth regulators in a strain of *Plutella xylostella* L. (Lepidoptera: Yponomeutidae) from Malaysia. *Pesticide Science* **33**, 359–370.

Jackson, K. and Webley, D. (1994) Effect of Dryacide on the physical properties of grains, pulses and oilseeds. In: *Proceedings of the 6th International Conference Stored Product Protection, Canberra, Australia, April 1994*, **2**, Highley, E., Wright, E.J., Banks, H.J., Champ, B.R. (eds), pp. 635–637.

Jay, E.G. (1986) Factors affecting the use of carbon dioxide for treating raw and processed agricultural products. *GASGA Seminar on Fumigation Technology in Developing Countries*. Tropical Development and Research Institute, London, pp 173–189.

Jutsum, A.R. and Gordon, R.F.S. (eds) (1985) *Insect Pheromones and Plant Protection*. John Wiley & Sons, Ltd, Chichester, UK.

Kader, A.A. and Ke, D. (1994) In: *Insect Pests and Fresh Horticultural Products: Treatments and Responses*. Paull, R.E. and Armstrong, J.W. (eds), CAB International, Wallingford, UK.

Katan, J. (1993) Replacing pesticides with nonchemical tools for the control of soilborne pathogens – a realistic goal? *Phytoparasitica* **21**, 95–99.

Kawada, H. and Hirano, M. (1996) Insecticidal effects of the insect growth regulators methoprene and pyriproxyfen on the cat flea (Siphonaptera: Pulicidae). *Journal of Medical Entomology* **33**, 819–822.

Kawakami, F. (1996) Import prohibited article and system of lifting import bans in Japan and procedures of disinfestation technology development tests. In: *Textbook for Vapour Heat Disinfestation Test Technicians*. Japan Fumigation Technology Association and Okinawa International Center, Naha, Okinawa, Japan International Cooperation Agency, Tokyo, pp. 10–13.

Keddis, M.E., Ayad, F.A., Abdel-Fattah, M.S. and El-Guindy, M.A. (1986) Studies of resistance to urea derivatives and their mixtures with insecticides in field strains of the cotton leaf-worm *Spodoptera littoralis* (Boisd.) during the cotton seasons 1983, 1984 and 1985. *Bulletin of the Entomological Society Egypt Economic Series*, **15**, 229–234.

Keiding, J., Jespersen, J.B. and El-Khodary, A.S. (1991) Resistance risk assessment of two insect development inhibitors, diflubenzuron and cryomazine, for control of the house fly *Musca domestica*. Part 1: larvicide tests with insecticide-resistant laboratory and Danish field populations. *Pesticide Science*, **32**, 187–206.

Kellen, W.R. and Hoffmann, D.F. (1987) Laboratory studies on the dissemination of a granulosis virus by healthy adults of the Indian meal moth, *Plodia interpunctella* (Lepidoptera: Pyralidae). *Environmental Entomology* **16**, 1231–1234.

Khatoon, N. and Heather, N.W. (1990) Susceptibility of *Dermestes maculatus* De Geer (Coleoptera: Dermestidae) to gamma radiation in a nitrogen atmosphere. *Journal of Stored Products Research* **26**, 227–232.

Knipling, E.F. (1955) Possibilities of insect control or eradication through the use of sexually sterile males. *Journal of Economic Entomology* **46**, 459–462.

Knulle, W. and Spadafora, R.R. (1969) Water vapour sorption and humidity relationship in *Liposcelis* (Insecta: Psocoptera). *Journal of Stored Products Research* **5**, 49–55.

Korunic, Z., Ormesher, P., Fields, P., White, N. and Cuperus, G. (1996) Diatomaceous earth an effective tool in integrated pest management. *Annual International Research Conference on Methyl Bromide Alternatives and Emissions Reduction, November 4–6, 1996*, Orlando, Florida, 81–82.

Kuwahara, Y., Fukami, H., Howard, R., Ishii, S., Matsumura, F. and Burkholder, W.E. (1978) Chemical studies on the Anobiidae: sex pheromone of the drugstore beetle, *Stegobium paniceum* (L.) (Coleoptera). *Tetrahedron* **34**, 1769–1774.

Kuwahara, Y., Thi My Yen, L., Tominaga, Y., Matsumoto, K. and Wada, Y. (1982) 1, 3, 5, 7, Tetramethyldecyl formate, lardolure: aggregation pheromone of the acarid mite, *Lardoglyphus konoi* (Sasa et Asanuma). *Agricultural and Biological Chemistry* **46**, p. 2283.

Labowsky, H.J. (1990) Soil disinfection in green houses and on open land. *Landtechnik* **45**, 270–271.

Landolt, P.J. and Heath, R.R. (1990) Attraction of female cabbage looper moths (Lepidoptera: Noctuidae) to male-produced sex pheromone. *Annals of the Entomological Society of America* **82**, 520–525.

Landolt, P.J. and Phillips, T.W. (1997) Host plant influences on sex pheromone behaviour of phytophagous insects. *Annual Review of Entomology* **42**, 371–391.

Le Torc'h, J-M. and Fleurat-Lessard, F. (1991) The effects of high pressure on insecticidal efficiency of modified atmospheres against *Sitophilus granarius* (L.) and *S. oryzae* (L.) (Coleoptera: Curculionidae). In: *Proceedings of the 5th International Working Conference on Stored Product Protection, September 1990, Bordeaux, France*, Fleurat-Lessard, F. and Ducom, P. (eds), vol. II, pp. 847–856.

Lewis, W.J. and Martin, W.R. Jr (1990) Semiochemicals for use with parasitoids: status and future. *Journal of Chemical Ecology* **16**, 3067–3089.

Lin, J.G., Hung, C.F. and Sun, C.N. (1989) Teflubenzuron resistance and microsomal mono-oxygenases in larvae of the diamondback moth. *Pesticide Biochemistry and Physiology* **35**, 20–25.

Lindberg, J.E. and Sorenson, E.I. (1959) Relationship between critical kernel temperature and moisture contents with respect to germinating properties of wheat. (In Swedish) *Kungliga Skogsoch Lantbruksakademiens Tidskrift Supplement 1*.

Lindquist, A.W. (1955) The use of gamma radiation for control or eradication of the screw-worm. *Journal of Economic Entomology* **48**, 467–469.

Lindquist, A.W. and Knipling, E.F. (1957) Recent advances in veterinary entomology. *Annual Review of Entomology* **2**, 181–202.

Locatelli, D.P. and Daolio, E. (1993) Effectiveness of carbon dioxide under reduced pressure against some insects infesting packaged rice. *Journal of Stored Products Research* **29**, 81–87.

López-Robles, J., Otto, A.A. and Hague, N.G.M. (1997) Evaluation of entomopathogenic nematodes on the beet cyst nematode *Heterodera schachtii*. *Annals of Applied Biology* **128**, 100–101.

McGaughey, W.H. (1985) Evaluation of *Bacillus thuringiensis* for controlling Indian meal moths in farm grain bins and elevator silos. *Journal of Economic Entomology* **78**, 1089–1094.

McGaughey, W.H. (1994) Problems of insect resistance to *Bacillus thuringiensis*. *Agriculture, Ecosystems and Environment* **49**, 95–102.

McLaughlin, A. (1994) Laboratory trials on desiccant dust insecticides. In: *Stored Product Protection*, Highley, E., Wright, E.J., Banks, H.J. and Champ, B.R. (eds) *Proceedings 6th International Conference of Stored Product-Protection*, Canberra, Australia, April 1994 **2**, CAB International, Wallingford, UK, pp. 638–645.

Maier, D.E., Rulon, R.A. and Mason, L.J. (1997) Chilled versus ambient aeration and fumigation of stored popcorn, Part 1: temperature management. *Journal of Stored Products Research* **33**, 39–49.

Malek, M.A. and Parveen, B. (1996) Natural products for managing *Tribolium* spp. (Coleoptera: Tenebrionidae). *Agricultural Zoology Reviews* **7**, 217–246.

Mallis, A. (1982) *Handbook of Pest Control*. MacNair-Dorland Company, New York, 1101 pp.

Mann, D.D., Jayas, D.S., White, N.D.G. and Muir, W.E. (1997) Sealing of welded-steel hopper bins for fumigation with carbon dioxide. *Canadian Agricultural Engineering* **39**, 91–97.

Manzelli, M.A. (1979) Controlling insect pests of stored tobacco by reducing their reproductive potentials. In: *Proceedings of the 2nd International Working Conference on Stored Product Entomology, Ibadan, Nigeria, 1978*, pp. 432–436.

Mayer, M.S. and McLaughlin, J.R. Jr (1991) *Handbook of Insect Pheromones and Sex Attractants*, CRC Press, Boca Raton, Florida.

MBTOC (1998) 1998 *Report of the Methyl Bromide Technical Options Committee. Assessment of Alternatives to Methyl Bromide, United Nations Environment Programme*, UNEP, Nairobi, Kenya.

Metcalf, R.L. and Luckmann, W.H. (1982) *Introduction to Pest Management* (2nd edn) John Wiley & Sons, Ltd.

Milner, R.J. (1972) *Nosema whitei*, a microsporidian pathogen of some species of *Tribolium*. III. Effect on *T. castaneum*. *Journal of Invertebrate Pathology*, **19**, 248–255.

Mitsuda, H., Kawai, F., Kuga, M. and Yamamoto, A. (1973) Mechanisms of carbon dioxide gas adsorption by grains and its application to skin packaging. *Journal of Nutrition Science* **19**, 71–83.

Moffitt, H.R. and Burditt, A.K. (1989*a*) Effects of low temperatures on three embryonic stages of the codling moth (Lepidoptera: Tortricidae). *Journal of Economic Entomology* **82**, 1379–1381.

Moffitt, H.R. and Burditt, A.K. (1989*b*) Low-temperature storage as a postharvest treatment for codling moth (Lepidoptera: Tortricidae). *Journal of Economic Entomology* **82**, 1679–1681.

Moy, J.H. (ed.) (1985) *Radiation Disinfestation of Food and Agricultural Products. Proceedings of a Conference in Honolulu, 1983*. University of Hawaii at Manoa, Honolulu.

Moy, J.H. (1988) Irradiation: a substitute to chemical fumigation of food. *IAEA-TECDOC*, **452**, p. 17.

Mummigatti, S.G., Ragunathan, A.N. and Karanth, N.G.K. (1994) *Bacillus thuringiensis* variety *tenebrionis* (DSM-2803) in the control of coleopteran pests of stored wheat. In: *Stored Product Protection*, Highley, E., Wright, E.J., Banks, H.J. and Champ, B.R. (eds), *Proceedings of the 6th International Working Conference on Stored-Product Protection* **2**, CAB International, Wallingford, UK, pp. 1112–1115.

Nakano, O. and Botton, M. (1997) Alternatives to methyl bromide use in flowers and ornamental plants in Brazil. *Brazilian Meeting on Alternatives to Methyl Bromide in Agriculture*. In: Müller, J.J.V. (ed.) EPAGRI, Palestra, Florianópolis, Brazìl, pp. 156–162.

Nataredja, Y.C. and Hodges, R.J. (1990) Commercial experience of sealed storage of bag stacks in Indonesia. In: *Fumigation and Controlled Atmosphere Storage of Grain*, Champ, B.R., Highley, E. and Banks, H.J. (eds), ACIAR Proceedings No. 25, pp. 197–202.

Navarro, S. and Calderon, M. (1974) Exposure on *Ephestia cautella* (Wlk.) pupae to carbon dioxide concentrations at different relative humidities: the effect on adult emergence and loss in weight. *Journal of Stored Products Research* **10**, 237–241.

Navarro, S., Donahaye, E. and Calderon, M. (1969) Observations on prolonged grain storage with forced aeration. *Journal of Stored Products Research* **5**, 73–81.

Nawrot, J. and Harmatha, J. (1994) Natural products as antifeedants against stored product insects. *Postharvest News and Information* **5**, 17 N–21 N.

Nederpel, L. (1979) Soil sterilization and pasteurization. In: *Soil Disinfestation*. Mulder, D. (ed.), Developments in Agricultural and Managed-Forest Ecology, 6. Elsevier Scientific Publishing Company, p. 29.

Nelson, S.O. (1972) Possibilities of controlling stored grain insects with R.F. energy. *The Journal of Microwave Power*, **7**, 231–239.

Neven, L. (1995) Combination treatments of pome fruit for quarantine. *Annual International Research Conference on Methyl Bromide Alternatives and Emission Reductions, November 6–8 1995*, San Diego, California.

Newton, J., Abey-Koch, M. and Pinniger, D.B. (1996) Controlled atmosphere treatment of textile pests in antique curtains using nitrogen hypoxia – a case study. In: *Proceedings of the 2nd International Conference on Insect Pests in the Urban Environment, Cambridge, UK*, Wildey, K.B. (ed.), pp. 329–339. BPC Wheatons Ltd, Exeter, UK.

Nichols, A.A. and Leaver, C.W. (1966) Methods of examining damp grain at harvest and after sealed and open storage: changes in the microflora of damp grain during sealed storage. *Journal of Applied Bacteriology* **29**, 566–581.

Nickson, P.J., Desmarchelier, J.M. and Gibbs, P. (1994) Combination of cooling with surface application of Dryacide to control insects. In: Highley, E., Wright, E.J., Banks, H.J. and Champ, B.R. (eds), *Proceedings of the 6th International Conference on Stored-Product Protection, Canberra, Australia, April 1994*, **2**, 646–649.

Oberlander, H., Silhacek, D.L., Shaaya, E. and Ishaaya, I. (1997) Current status and future perspectives of the use of insect growth regulators for the control of stored product insects. *Journal of Stored Products Research* **33**, 1–6.

Ogawa, K. (1997) The key to success in mating disruption. In: *Technology Transfer in Mating Disruption. Proceedings of a Working Group Meeting in Montpellier, France, 9–10 September 1996.* Bulletin OILB/SROP **20**(1), 1, 1–9.

Phillips, T.W. (1994) Pheromones of stored-product insects: current status and future perspectives. In: *Stored Product Protection. Proceedings of the 6th International Working Conference on Stored Product Protection*, Canberra, Australia, E. Highley, E.J. Wright, H.J. Banks and B.R. Champ (eds), pp. 479–486. CAB International, Wallingford, UK.

Phillips, T.W. (1997) Semiochemicals of stored-product insects: research and applications. *Journal of Stored Products Research* **33**, 17–30.

Pierce, A.M., Pierce, H.D. Jr, Oehlschlager, A.C. and Borden, J.H. (1985) Macrolide aggregation pheromones in *Oryzaephilus surinamensis* and *O. mercator*. *Journal of Agricultural and Food Chemistry* **33**, 848–852.

Pimentel, D. (1991) *CRC Handbook of Pest Management*. (2nd edn) CRC Press, Boca Raton, Florida.

Pixton, S.W., Warburton, S. and Hill, S.T. (1975) Long term storage of wheat III. Some changes observed during 16 years storage. *Journal of Stored Product of Research* **11**, 177–185.

Press, J.W., Flaherty, B.R. and Arbogast, R.T. (1975) Control of the red flour beetle, *Tribolium castaneum* in a warehouse by a predaceous bug, *Xylocoris flavipes*. *Journal of the Georgia Entomological Society* **10**, 76–78.

Press, J.W., Cline, C.D. and Flaherty, B.R. (1982) A comparison of two parasitoids, *Bracon hebetor* and *Venturia canescens* and a predator, *Xylocoris flavipes* in suppressing residual populations of the almond moth, *Ephestia cautella*. *Journal of Kansas Entomological Society* **55**, 725–728.

Prozell, S., Reichmuth, C., Ziegleder, G., Schartmann, B., Matissek, R., Kraus, J., Gerard, D. and Rogg, S. (1997) Control of pests and quality aspects in cocoa beans and hazel nuts and diffusion experiments in compressed tobacco with carbon dioxide under high pressure. In: *Proceedings International Conference on Controlled Atmosphere and Fumigation in Stored Products, April, 1996*, Donahaye, E.J., Navarro, S. and Varnava, A. (eds) Printco Ltd, Nicosia, Cyprus, pp. 325–333.

Pulpan, J. and Verner, P.H. (1965) Control of tyroglyphid mites in stored grain by the predatory mite, *Cheyletus eruditus*. *Canadian Journal of Zoology* **43**, 417–432.

Ridgway, R.L., Silverstein, R.M. and Inscoe, M.N. (eds) (1990) *Behavior-Modifying Chemicals for Pest Management: Applications of Pheromones and Other Attractants.* Marcel Dekker, New York.

Robertson, A.R. and Baritelle, J.L. (1996) Economic feasibility of alternatives for postharvest disinfestation of California dried fruits and nuts. In: *1996 Annual International*

Research Conference on Methyl Bromide Alternatives and Emissions Reductions, 4–6 November 1996, Orlando, Florida, USA, p. IV.

Robinson, W. (1926) Low temperature and moisture as factors in the ecology of the rice weevil, *Sitophilus oryzae* L. and the granary weevil, *Sitophilus granarius* L. *University of Minnesota Exp. Station Report*, **41**, 40 pp.

Rodríguez-Kábana, R. (1998) Alternatives to methyl bromide (MB) soil fumigation. In: *Alternatives to Methyl Bromide for the Southern European Countries*. Bello, A., González, J.A., Arias, M. and Rodríguez-Kábana, R. (eds) DG XI, EU, CSIC, Madrid, pp. 17–34.

Roorda-van-Eysinga, J.P.N.L. (1984) Bromine in glasshouse lettuce as affected by chemical soil disinfectants and steam sterilization. *Acta Horticulturae* **154**, 262–268.

Runia, W.T. (1983) A recent development in steam sterilisation. *Acta Horticulturae* **152**, p. 195.

Rust, M.K. and Dryden, M.W. (1997) The biology, ecology and management of the cat flea. *Annual Review of Entomology* **42**, 451–473.

Samson, P.R., Parker, R.J. and Hall, E.A. (1990) Efficacy of the insect growth regulators methoprene, fenoxycarb and diflubenzuron against *Rhyzopertha dominica* (F.) (Coleoptera; Bostrichidae) on maize and paddy rice. *Journal of Stored Products Research* **26**, 215–221.

Sanders, C.J. (1997) Mechanisms of mating disruption in moths. In: *Insect Pheromone Research: New Directions*, Carde, R.T. and Minks, A.K. (eds), 1st International Symposium on Insect Pheromones, Wageningen, Netherlands, 6–11 March 1994, Chapman and Hall, New York.

Scherer, R. and Rakotonandrasana (1993) Barrier treatment with a benzoyl urea insect growth regulator against *Locusta migratoria capito* (Sauss) hopper bands in Madagascar. *International Journal of Pest Management* **39**, 411–417.

Schmuff, N., Phillips, J.K. Burkholder, W.E., Fales, H.M., Chen, C., Roller, P. and Ma, M. (1984) The chemical identification of the rice and maize weevil pheromones. *Tetrahedron Letters* **25**, 1533–1534.

Scholler, M. and Prozell, S. (1996) Response of *Trichogramma evanescens* to synthetic (Z, E)-9, 12-tetra-decadienyl acetate (TDA), a sex pheromone component of *Ephestia kuehniella* and *Plodia interpunctella* (Hymenoptera: Trichogrammatidae – Lepidoptera: Pyralidae). In: *Proceedings of the 20th International Congress on Entomology*, p. 646, Florence, Italy.

Scholler, M., Prozell, S., Al-Kirshi, A.-G. and Reichmuth, Ch. (1997) Towards biological control as a major component of integrated pest management in stored product protection. *Journal of Stored Products Research* **33**, 81–97.

Scopes, N. and Stables, L. (1989) *Pest and Disease Control Handbook*. (3rd edn) BCPC.

Seo, S.T., Chambers, D.L., Komura, M. and Lee, C.V.L. (1970) Mortality of mango weevils (Coleoptera: Curculionidae) in mangos treated by dielectric heating. *Journal of Economic Entomology* **63**, 1977–1978.

Shapas, T., Burkholder, W.E. and Boush, G.M. (1977) Population suppression of *Trogoderma glabrum* using pheromone luring for protozoan pathogen dissemination. *Journal of Economic Entomology* **70**, 469–474.

Sharp, J.L. and Spalding, D.H. (1984) Hot water as a quarantine treatment for Florida mangoes infested with Caribbean fruit fly. *Proceedings of the Florida State Horticultural Society* **97**, 355–357.

Shazali, M.E.H. and Smith, R.H. (1986) Life history studies of externally feeding pests of stored sorghum: *Corcyra cephalonica* (Stamp) and *Tribolium castaneum* (Hbst.) *Journal of Stored Product Research* **22**, 55–61.

Siddig, S.A. (1980) Efficacy and persistence of powdered neem seeds for treatment of stored wheat against *Trogoderma granarium*. In: *Proceedings of the 1st International*

Neem Conference, Rothach-Egern, Federal Republic of Germany, Schmutterer, H., Ascher, K.R.S. and Rembold, H. (eds), pp. 251–257.

Smagghe, G., Salem, H., Tirry, L. and Degheele, D. (1996) Action of a novel insect growth regulator, tebufenozide against different development stages of four stored product insects. *Parasitica* **52**, 61–69.

Smith, L.B. (1970) Effects of cold acclimation on supercooling and survival of the rusty grain beetle, *Cryptolestes ferrugineus* (Stephens) (Coleoptera: Cucujidae) at sub-zero temperatures. *Canadian Journal of Zoology* **48**, 853–858.

Smith, L. (1994) Computer simulation model for biological control of maize weevil by the parasitoids, *Anisopteromalus calandrae*. *Proceedings of the 6th International Working Conference on Stored Product Protection*, **2**, 1147–1151.

Sonneveld, C. (1979) Changes in chemical properties of soil caused by steam sterilization. In: *Soil Disinfestation. Developments in Agricultural and Managed-Forest Ecology*, **6**. Mulder, D. (ed.), Elsevier Scientific Publishing Company, p. 39.

Staal, G.B. (1975) Insect growth regulators with juvenile hormone activity. *Annual Review of Entomology* **20**, 417–460.

Staryzk, J.R. (1996) Utilization of pheromones for the prognosis and control of secondary insect pests in mountain forests. *Sylvan* **140**, 23–36.

Stirling, G.R., Dullahide, S.R. and Nikulin A. (1995) Management of lesion nematode (*Pratylenchus jordannensis*) on replant apple trees. *Australian Journal of Experimental Agriculture* **35**, 247–258.

Su Nan Yao, Scheffrahn, R.H. and Su, N.Y. (1998) A review of subterranean termite control practices and prospects for integrated pest management programmes. *Integrated Pest Management Reviews* **3**, 1–13.

Sutherland, J.W. (1968) Control of insects in a wheat store with an experimental aeration system. *Journal of Agricultural Engineering Research* **13**, 210–219.

Sutherland, J.W., Evans, D.E., Fane, A.G. and Thorpe, G.R. (1987) Disinfestation of grain with heated air. *Proceedings of the 4th International Working Conference on Stored-Product Protection*, Tel Aviv, Israel, pp. 261–274.

Suzuki, T. (1980) 4,8-dimethyldecanal: the aggregation pheromones of the flour beetles *Tribolium castaneum* and *T. confusum* (Coleoptera: Tenebrionidae). *Agricultural and Biological Chemistry* **44**, 2519–2520.

Thorpe, G.R. and Evans, D.E. (1983) *The Drying and Disinfestation of Cereal Grains*. Australian Institute of Energy, Symposium on Drying and Curing, Melbourne.

Thuy, P.T., Dien, L.D. and Van, N.G. (1994) Research on multiplication of *Beauveria bassiana* fungus and preliminary utilisation of Bb bioproduct for pest management in stored products in Vietnam. *Proceedings of the 6th International Working Conference Stored Product Protection* **2**, 1132–1133.

Tilton, E.W. and Brower, J.H. (1983) Radiation effects on arthropods. In: *Preserving of Food by Ionizing Radiation*, eds Josephson, E.S. and Peterson, M.S. Vol. 2, 269–316. CRC Press, Boca Raton, Florida.

Tilton, E.W. and Brower, J.H. (1987) Ionizing radiation for insect control in grain and grain products. *Cereal Food World* **32**, 330–335.

Trematerra, P. (1994) The use of sex pheromones to control *Ephestia kuehniella* Zeller (Mediterranean flour moth) in flour mills by mass trapping and attracticide (lure and kill) methods. In: *Stored Product Protection. Proceedings of the 6th International Working Conference on Stored Product Protection, Canberra, Australia*, Highley, E., Wright, E.J., Banks, H.J. and Champ, B.R. (eds), pp. 375–382. CAB International, Wallingford, UK.

Tzortzakakis, E.A. and Gowen, S.R. (1994) Evaluation of *Pasteuria penetrans* alone and in combination with oxamyl, plant resistance and solarization for control of *Meloidogyne* spp. on vegetables grown in greenhouses in Crete. *Crop Protection* **13**, 455–462.

USDA (1989) Use of irradiation as a quarantine treatment for fresh fruit of papaya from Hawaii. *United States Department of Agriculture. Federal Register* **54**, 387–393.

USFDA (1986) Irradiation in the production, processing and handling of food. Final rule. *United States Food and Drug Administration. Federal Register* **51**, 13 376–13 399.
Williams, C.M. (1967) Third generation pesticides. *Scientific American* **217**, 13–17.
Williams, H.J., Silverstein, R.M., Burkholder, W.E. and Khorramshahi, A. (1981) Dominicalure 1 and 2: components of the aggregation pheromone from male lesser grain borer *Rhyzopertha dominica* (F.) *Journal of Chemical Ecology* **7**, 759–780.
Williams, L.A.D., Ajai Mansingh and Mansingh, A. (1996) The insecticidal and acaricidal actions of compounds from *Azadirachta indica* (A. Juss.) and their use in tropical pest management. *Integrated Pest Management Reviews* **1**, 133–145.
Williams, R.N. and Floyd, E.H. (1971) Effect of two parasites, *Anisopteromalus calandrae* and *Choetospila elegans*, upon populations of the maize weevil, under laboratory and natural conditions. *Journal of Economic Entomology* **64**, 1407–1408.
Williamson, M.R. and Winkelman, P.M. (1989) *1989 International Winter Meeting, American Society of Agricultural Engineers, Quebec, Canada*. ASAE, St Joseph, Missouri.
Willis, M.A. and Birch, M.C. (1982) Male lek formation and female calling in a population of the arctiid moth, *Estigmene acrea*. *Science* **218**, 168–170.
Wong, L. (1998) *Communicating Food Safety Technologies*. Presentation to the Science and Public Policy Institute, April 7 1998, Washington DC.
Wooten, M. (1985) Application of gamma irradiation to cereals and cereal products. In: *Regional Workshop on Commercialisation of Ionising Energy Treatment of Food, Lucas Heights, 1985, Lecture 14*. Australian Atomic Energy Commission, Lucas Heights, New South Wales.
Yokoyama, V.Y. and Miller, G.T. (1989) Response of codling moth and oriental fruit moth (Lepidoptera: Tortricidae) immatures to low-temperature storage of stone fruits. *Journal of Economic Entomology* **82**, p. 1152.
Zakladnoi, G.A., Men'shenin A.I., Pertsovskii, E.S., Salinov, R.A., Cherepkov, V.G. and Krssheminskii, V.S. (1982) Industrial application of radiation deinsectification of grain. *Soviet Atomic Energy* **52**, 74–78.
Zdarkova, E. and Horak, E. (1990) Preventative biological control of stored food mites in empty stores using *Cheyletus eruditus*. *Crop Protection* **9**, 378–382.

INDEX

Optimising Pesticide Use Edited by M. Wilson
 ISBN: 0-471-49075-X

With kind thanks to Geraldine Begley for compilation of this index.